MW00856329

Uncovering Dinosaur Behavior

Uncovering Dinosaur Behavior

What They Did and How We Know

David Hone

SELECT ILLUSTRATIONS BY GABRIEL UGUETO

PRINCETON UNIVERSITY PRESS   PRINCETON & OXFORD

Copyright © 2024 by David Hone
Illustrations copyright © 2024 by Gabriel Ugueto

Princeton University Press is committed to the protection of copyright and the intellectual property our authors entrust to us. Copyright promotes the progress and integrity of knowledge. Thank you for supporting free speech and the global exchange of ideas by purchasing an authorized edition of this book. If you wish to reproduce or distribute any part of it in any form, please obtain permission.

Requests for permission to reproduce material from this work should be sent to permissions@press.princeton.edu

Published by Princeton University Press
41 William Street, Princeton, New Jersey 08540
99 Banbury Road, Oxford OX2 6JX

press.princeton.edu

All Rights Reserved

Library of Congress Cataloging-in-Publication Data

Names: Hone, David (Paleontologist), author.
Title: Uncovering dinosaur behavior : what they did and how we know / David Hone.
Description: Princeton : Princeton University Press, [2024] | Includes bibliographical references and index.
Identifiers: LCCN 2024012620 (print) | LCCN 2024012621 (ebook) | ISBN 9780691215914 (hardback) | ISBN 9780691255330 (ebook)
Subjects: LCSH: Dinosaurs—Behavior. | Dinosaurs—Behavior—Evolution. | Dinosaurs—Ecology. | BISAC: SCIENCE / Paleontology | SCIENCE / Life Sciences / Zoology / Ethology (Animal Behavior)
Classification: LCC QE861.6.B44 H66 2024 (print) | LCC QE861.6.B44 (ebook) | DDC 567.9—dc23/eng/20240409
LC record available at https://lccn.loc.gov/2024012620
LC ebook record available at https://lccn.loc.gov/2024012621

British Library Cataloging-in-Publication Data is available

Editorial: Alison Kalett and Hallie Schaeffer
Production Editorial: Kathleen Cioffi
Production: Jacquie Poirier
Publicity: Matthew Taylor and Kate Farquhar-Thomson

Jacket illustration: A group of the Late Triassic theropod *Coelophysis* engaging in some of the huge variety of activities of which dinosaurs were capable. Artwork by Gabriel Ugueto

This book has been composed in Signifier and The Future

Printed in the United States of America

10 9 8 7 6 5 4 3 2 1

To Max, without whom this book would never have come to fruition

CONTENTS

For almost as long as scientists have been unearthing the fossils of dinosaurs, there has been speculation, discussion, and study of their possible behaviors. This started with the simplest of inferences—such as "Those animals with leaf-shaped teeth bearing large serrations were herbivorous"—and soon moved on to considerations about dinosaurs living in groups, climbing trees, hunting for food, and everything else that goes into the behavioral repertoire of a large terrestrial vertebrate. In the mid-1800s, dinosaur bones mostly amounted to a relative handful of very partial skeletons, and animal behavior was not even a field of zoological study in its own right, so one can easily forgive some extrapolations that are, in glorious hindsight, somewhat absurd.

Still, the fundamental challenge that plagued these early paleontologists is still with us. If one would care to study, say, the problem-solving abilities of a tortoise, the memory of a crow, the preferred group size of an iguana, or the mating rituals of a parrot, one has only to observe the living animals to gain some insights into their lives. We can attach trackers to everything from whales to bees to see when and where they forage and for how long, and see how this varies with temperature, latitude, season, and however many of the other infinite variables we care to assess. Single animals can be followed for their entire lives, and multiple generations can be studied in the lab, their genomes analyzed, relationships determined, and experiments set up to test ideas about how and why they respond to certain stimuli. Quite simply, none of this is available to the paleoethologist, so everything about the lives of ancient species must be inferred, often from very limited evidence, and therefore always with at least a grain (and sometimes a full cellar) of salt.

That is not to say that the study of dinosaur behavior is simply speculation or has no foundation in science, or that there is no way through the often conflicting positions that have been advocated for one behavior over another. Ideas can be tested and supported or rejected, comparisons made to living species and their patterns of behavior, and mechanical hypotheses (is this tooth strong enough to bite through that bone?) rigorously assessed. There are (big) gaps and uncertainties, and some previously confidently held understandings need tearing down or at least a rethink, but recent years have brought a reassessment of dinosaur behavior, and it is no longer a case of "anything goes"—throwing out some unsupported and wild hypothesis in the last line of a paper and leaving it there. Well, it's rather less common, at least. But it does mean we have more, and more thoughtful and insightful, pieces being put forward by more researchers on more new fossils. It is therefore an exciting time for anyone with an interest in dinosaur behavior.

The intent of this book is to be one for paleontologists who want to know about behavior, and a book for ethologists who want to know something about dinosaurs, and hopefully something of interest to biologists in general. The obvious risk, of course, is that it will fall short in every area and be a book for no one, but I can't ignore that, for the early chapters in particular, it will be necessary to cover some things that may well be important new information for one reader and oversimplified or trivial for another. On a related note, while the content is hopefully rigorous in its approach to the scientific literature, my writing style tends to be more conversational in tone, but is at least intended to be both accessible and useful to the lay reader, the student, and the academic. Whether or not I have succeeded in these aims, only time (and the online reviews) will tell.

While the book has been written with the expectation that it would be read through, I am well aware that many people will want to read only specific chapters or sections. Given that it is often hard to separate out many features of behaviors from their intimate association with other aspects of biology (e.g., horns that were used to fight with rivals were also likely to be deployed as visual signals), some features make multiple appearances in the text, and there is a certain amount of cross-referencing within the book to hopefully make it easier for the reader to track down topics that occur in more than one chapter, or where areas of research intersect.

Although scientific studies that specifically look at dinosaur behavior are relatively few and far between compared to, say, studies
of their anatomy and evolutionary history, there are nevertheless a
considerable number of dedicated papers and book chapters in this
area, and plenty more that comment on key points in passing. It
would be impossible to curate an entirely comprehensive approach
to this material, and so, in my choice of citations, I have chosen to
focus on more recent references and overarching reviews or classic
and important papers. Similarly, there are more than a few fringe
hypotheses or ideas out there for dinosaurs—more than for most or
any other fossil groups—and attention cannot be given to every single
known suggestion for, say, the function of the arms of *Tyrannosaurus*.
I will therefore generally stick to ideas that have been relatively well
supported in the literature, or at least those that have been more
prominent in discussions of dinosaur behavior.

Moreover, this book also takes what is perhaps a liberal approach
to behavior, and includes aspects of animal biology such as their occupation of different habitats and the nature of their senses. There
might be a lack of detail about how a dinosaur that lived in a coastal
region versus one that inhabited upland woodlands would have in-
teracted with its environment, but this aspect of their ecology would
have profoundly influenced each species' behavior, given differences
in climate and available cover, for example. Similarly, while any spe-
cifics about their behavior might be hard to determine, working out
that a species had high visual acuity and limited olfaction, for example, would clearly also be important considerations for what kind of
stimuli they might respond to and what was ignored. If this book
stuck only to direct interactions with other animals or direct evidence of behavior, it would be considerably shorter. Instead, the major
chapters attempt to cover both the background biology to various
aspects of behavior (why do animals live in groups? why do animals
fight?) and then also hypotheses and available data that have been
established for various behaviors in various dinosaurs. So rather than
a list of studies that show that species X likely did Y, this is hopefully
a more integrated work that covers our general understanding of
dinosaur behavior, and by looking at the origins of hypotheses and

the mechanisms of obtaining these conclusions, provides a basis for further inferences and studies in the future. Therefore, this is not (just) a catalogue of what we know (or what we are arguing about), but an account of how we got there, and perhaps routes to resolve the conflicts and develop more knowledge going forward.

It is also the case that some areas of research into dinosaur behavior are very deep and others much more limited, for various reasons (inevitably, large carnivores tend to attract a lot of interest, and in contrast the giant sauropods have very few preserved skulls, making their brains and senses hard to study), and I'll be unapologetic in

FIGURE 0.1 The modern study of dinosaur behavior is a case of trying to reconstruct that which is lost based on what we can take from the limited data available in the fossil record and from our understanding of living animals. Here, this is exemplified by an extinct and an extant display structure: a *Triceratops* frill and a peacock train. Photograph courtesy of Gareth Monger.

following those patterns here. This means that some major groups of dinosaurs will get very little attention, and the overexposed and hypercharismatic megafauna will continue to dominate. However, given how many studies do focus on single species, so that extrapolation or inference is necessary to say anything meaningful about their relatives, this is not such a bad choice as it may first appear. There may arguably be too many papers focusing on *Tyrannosaurus rex*, for example, but it does provide an excellent point of reference for its closest relatives, and to a certain degree for large predatory dinosaurs in general. Thus this book will reflect the current literature and research trends, even if that's not always ideal, and I can at least flag for those interested in the field that there is a clear need to redress some balances in the future.

ACKNOWLEDGMENTS

Over the years I've had innumerable conversations with colleagues and collaborators about the behavior of dinosaurs. Many have been enormously productive, with few being frustrating, and generally disagreements did not result in anyone needing to visit the hospital. These have, though, shaped my thoughts on and understanding of dinosaur behavior, and whether we worked together on a research paper, exchanged messages by email, or had a not-entirely-sober conversation in the bar, I wanted to try and thank them and acknowledge their contributions to what ultimately became this book.

So my great thanks to Victoria Arbour, Chris Barker, Paul Barrett, Mike Benton, Don Brinkman, Caleb Brown, Steve Brusatte, David Button, Thomas Carr, Andrea Cau, Jonah Choiniere, Dan Chure, Jim Clarke, Jennifer Colbourne, John Conway, Phil Currie, Innes Cuthill, Alex Dececchi, Brian Engh, Doug Emlen, David Evans, Peter Falkingham, Andy Farke, Chris Faulkes, Adam Fitch, Kiersten Formoso, Cathy Forster, Zhang Fucheng, Mike Habib, Yara Haridy, Scott Hartman, Don Henderson, Tom Holtz, ReBecca Hunt-Foster, John Hutchinson, Jim Kirkland, Andrew Knapp, Rob Knell, Martin Kundrát, Stephan Lautenschlager, Iszi Lawrence, Susie Maidment, Heinrich Mallison, Jordan Mallon, Julia McHugh, Andrew Milner, Karen Moreno, Cass Morrison, Ali Nabavizadeh, Darren Naish, Bob Nicholls, Mark Norell, Scott Persons, Oliver Rauhut, Emily Rayfield, Luis Rey, Héctor Rivera-Sylva, Manabu Sakamoto, Dave Shuker, Eric Snively, Corwin Sullivan, Darren Tanke, Mike Taylor, Emmanuel Tschopp, Gabriel Ugueto, Paul Upchurch, Mahito Watabe, Matt Wedel, Larry Witmer, Mark Witton, Cary Woodruff, Xing Xu, Adam Yates, and plenty more that I am sure I have forgotten, and to whom I apologize in advance.

I want to thank my agent Max Edwards and the people at Aevitas Creative for representing me and helping get this book over the line. My thanks too to Alison Kalett and Hallie Schaeffer at Princeton

University Press, who commissioned the book and helped create and shape the work into what it is now.

This would not be the book it is without the artistry of Gabriel Ugueto in recreating these animals and their world, and more importantly depicting their behaviors in a way that's both interesting and as accurate as possible. I can't thank him enough for his efforts.

Similarly, I must thank all those who kindly handed over or arranged for me to be able to get photos to illustrate and supplement the artwork, and so thanks to Raimund Albersdörfer, Chris Barker, Danny Barta, Michael Buckland, Caleb Brown, Adam Fitch, Phoebe Griffith, ReBecca Hunt-Foster, John Hutchinson, Marc Jones, Rob Knell, Martin Kundrát, Bruce Lauer, René Lauer, Stephan Lautenschlager, Skye McDavid, Jason McDonald, Colin McHenry, Eric Metz, Gareth Monger, Ali Nabavizadeh, Rodrigo Pellegrini, Irene Proebsting, Cristiano Dal Sasso, Corwin Sullivan, Neralie Thorp, Paolo Viscardi, Cary Woodruff, and Jay Young.

Finally, thanks to Andy Farke, Dave Shuker, and Cary Woodruff and two anonymous referees for kindly reading through an early version of the whole book to offer their thoughts and comments to improve it. I should, of course, say that any remaining errors or omissions are technically mine, but really I blame them for not spotting these earlier.

CHAPTER 1

An Introduction to Dinosaurs

The Dinosauria represent an extremely successful group of tetrapods that were the dominant terrestrial group of most of the Mesozoic Era, and in the birds, have comfortably over 10,000 living species as descendants. (Throughout this book I will refer to dinosaurs and Dinosauria as a paraphyletic group, therefore excluding both Mesozoic and modern birds unless explicitly stated otherwise.) Although various fragmentary fossils that we now recognize as being dinosaurian were being discussed in the eighteenth century, they came to prominence with the naming of the British carnivore *Megalosaurus* in 1824[1] and herbivore *Iguanodon* in 1825[2] and the coining of the name "Dinosauria," or "fearfully great lizards," by Richard Owen in 1842.[3] The dinosaurs quickly grew in both numbers of species described and taxonomic ranks recognized, with ever more, ever better, and ever larger fossils being discovered. Dinosaurs soon became established as a major area of interest in the burgeoning field of paleontology and have become central to the study of the history of life on Earth (figure 1.1).

Dinosaurs are no longer considered the cold-blooded, tail-dragging, stupid, lizard-like monsters of the Victorian age, but are instead recognized as animals that were upright, active, fast-growing, and if not especially intelligent, certainly not stupid. Fossils of dinosaurs are now known from dozens of countries and from every continent, including Antarctica;[4] in life they occupied every ecosystem from mountains to deltas and deserts to forests,[5] and they included in their number the largest terrestrial animals of all time.[6] We have fossils of dinosaurs with their skin and feathers intact,[7] as well as other soft-tissue structures like cockscomb head crests,[8] and even traces of the original patterns and colors of the living animals.[9]

FIGURE 1.1 There is perhaps no such thing as a "typical" dinosaur given their huge range in shape, size, and, undoubtedly, behavior. Here, at least, is an exemplar: the large Late Jurassic theropod *Allosaurus*, by far the most common carnivore in its ecosystem, and both well represented in the fossil record and well studied. Shown are a restored skeleton and a life reconstruction of the animal. Artwork by Gabriel Ugueto.

These discoveries, coupled with two centuries of research, have enabled huge advances in the reconstruction of every aspect of the biology of dinosaurs. These ancient organisms are now firmly established in modern science, though there remain some large gaps and areas of uncertainty in our understanding of these incredible animals.

Origins and Relationships

Dinosaurs are members of the reptilian clade Archosauria that includes modern crocodilians (and their extinct ancestors and relatives), the Mesozoic flying reptiles the pterosaurs, and a number of other groups. (It is increasingly likely that the chelonians—turtles, terrapins, and tortoises—are also archosaurs or their closest relatives, though this is still a subject of academic debate and is not currently certain; for simplicity here they will be excluded from this clade.) The archosaurs are united in the presence of an antorbital fenestra (an opening in the skull between the naris and orbit, though this is secondarily lost in many groups), serrated teeth set in sockets in the jaws, and an upright stance with the limbs held under the body (though as with so many defining features, evolutionary history has modified these in various groups, most notably to give the semi-sprawling posture of modern crocodilians).[10]

Among these various archosaurs were a group of small (under 2 m in total length), bipedal carnivores or omnivores called the dinosauromorphs that ultimately gave rise to the dinosaurs. Sometime in the late part of the Middle Triassic around 240 million years ago,[11] the dinosaurs split from their dinosauromorph ancestors. At this time, there was a single major landmass, Pangea, that was largely hot and dry, although early dinosaurs may have favored the colder regions of this.[12] Inevitably, the early dinosaurs look extremely similar to their nearest relatives, and the exact point of separation and differentiation between the two is uncertain. The genus *Nyasasaurus* from the Middle Triassic of Tanzania, for example, may be either the oldest known dinosaur or the nearest dinosauromorph relative to the dinosaurs,[11] such is the closeness between the two at this point.

Early dinosaurs were small, bipedal carnivores. They were a relatively minor component of the early Late Triassic terrestrial ecosystems; though they diversified and grew in size, it was not until the extinction at the end of the Late Triassic that dinosaurs became the dominant terrestrial group of the Mesozoic.[12] The dinosaurs have long been split into three major clades: the theropods, which were bipedal and predominantly carnivorous and are the ancestors of birds; the sauropodomorphs, which were herbivorous and were long-necked and mostly large; and the ornithischians, which were herbivorous and produced a much greater diversity of body forms than the other groups.[10] The theropods and sauropodomorphs are united into the Saurischia ("lizard-hipped reptiles" because of their anteriorly directed pubis, though derived theropods and birds reverse this) as the sibling taxon to the Ornithischia (figure 1.2), though recently this interpretation has been challenged. Reinterpretation of

FIGURE 1.2 The major lineages of the dinosaurs, with Theropoda exemplified by *Tyrannosaurus* (top), Sauropodomorpha by *Diplodocus* (bottom), and Ornithischia by *Stegosaurus* (center) and *Triceratops* (right). Artwork by Gabriel Ugueto.

a number of traits coupled with new discoveries have suggested that in fact the theropods may lie with the ornithischians at the expense of the sauropodomorphs.[13] This is controversial, though certainly possible given the apparent absence of ornithischians in the Triassic (although see [14]). The exact nature of this split is largely irrelevant from the point of view of discussions of behavior, though for simplicity and clarity the "traditional" split into Saurischia and Ornithischia is used throughout.

Major Groups

Dinosaurs are known from around 1,500 valid genera, though this number is currently growing by 30 to 50 genera each year and has been for at least a decade. Each of the three main clades is well represented with several hundred taxa, though the theropods are the most numerous of these and the sauropodomorphs the least.

The theropods range from small animals that were perhaps as little as one kilo through to giants like the largest tyrannosaurs that were over 13 m long and weighed perhaps 7 tons or more. Although the basic bipedal form of theropods was essentially universal, they vary enormously in skull size and in the lengths of the neck, legs, and especially the arms.[15] Ancestrally, theropods were carnivorous, with sharp claws on the hands and feet, and all known truly carnivorous dinosaurs are theropods, although some were specialist fish or insect eaters, and members of a number of lineages in the Jurassic and Cretaceous switched to omnivory or even herbivory at various times.[16] Numerous derived theropods are preserved with feathers and also show various birdlike features, such as hollow bones (part of the system of extension of the pulmonary tracts termed air sacs), and extensive research demonstrates that the birds ultimately derived from a group of small theropods in the Middle Jurassic.[17] In particular, numerous small and feathered theropods are known from fossil beds of exceptional preservation, and so the transition to birds is one of the best studied and understood major evolutionary transitions in biology.

The sauropodomorphs are famous for producing the largest terrestrial animals of all time. In the Triassic, the early sauropodomorphs (often termed "prosauropods") were large for terrestrial animals at the time, but were small by the standards of those that came later.[6] The prosauropods were predominantly bipedal (though some were quadrupeds as juveniles) and were characterized by relatively long necks with small heads and a large body, all adaptations for herbivory,[6] though a handful of the very earliest species were probably omnivorous. In the Jurassic, these were replaced by the sauropods, quadrupedal animals that retained the small head, but now typically on an even longer neck,[18] and would go on to produce giants that could exceed 50 tons. Sauropodomorphs also had pneumatic vertebrae invaded by air sacs, and the largest of these animals had numerous vertebrae that were very light considering their size. As a consequence of this, however, large parts of their skeletons are often poorly preserved, although the apneumatic robust and massive limb bones are often found. Sauropods dominated the Jurassic landscapes, though they were rather less common in the Cretaceous, especially in the northern hemisphere.

It would mischaracterize the ornithischians to call them simply "all the other dinosaurs," though they do show considerable variation in general form, unlike the two saurischian lineages. The ornithischians range from small, 1-meter-long bipedal animals up to 15-meter-long giant quadrupeds; there were animals covered in armored plates and spikes, those with giant frills, crests, bosses, and horns on their heads (as well as plenty that were unadorned); and at least some also had feather-like filaments on the body in addition to scales.[19] Again, with the exception of perhaps a few of the earliest taxa which may have been omnivores, these were all herbivorous animals. Notably, we see some extraordinary adaptations to consume and process plants in the ornithischians (compared to the sauropodomorphs, who largely were bulk feeders on as much plant matter as possible), and this included the evolution of a beak at the front of the jaw with teeth behind.[16] Although rare early on, and currently absent in the Triassic, the ornithischians were important components of Jurassic faunas, and in the Cretaceous became the dominant herbivores in most terrestrial ecosystems.

For more information on the major clades within the Theropoda, Sauropodomorpha and Ornithischia see the guide in the Appendix (page 159).

Basic Biology

Although there are some serious gaps in the fossil record for dinosaurs, they appear to have occupied every major terrestrial ecosystem on Earth given their truly global distribution,[4] and this includes everything from sandy deserts[20] to Arctic ice.[21] As with many large modern animals, individual species are found across multiple environments, and there is strong evidence for migration of various dinosaurs;[22] annual migrations would have been the norm for at least some species.

As noted above, the dinosaurs include herbivores, omnivores, and carnivores,[23] but beyond this most general of statements about their feeding ecology, we have strong evidence for both generalist and specialist diets. Among herbivores, there are taxa known to have been selective or bulk feeders, and those that were high or low browsers. Among various theropods, there is evidence for long-distance pursuit predation, carnivory of other large animals, scavenging, piscivory, and insectivory. Evidence for these various patterns of behavior come from numerous sources, such as fossilized coprolites (feces) and regurgitated pellets, fossil stomach contents, tooth shape and wear on teeth, head shape and mouth sizes, and bite traces on bones, among others.[24]

The dinosaur fossil record is sufficient to trace some major evolutionary patterns through their history, and as such we have been able to identify important trends across their 180 million years of evolution. Notably, dinosaurs tended to get bigger over time, producing numerous large lineages, though the transition to birds came from a sustained reduction in size of theropods across tens of millions of years.[12] One other important transition is the repeated shift from bipedality to quadrupediality in various dinosaur lineages,[25] something that is notably seen in the various herbivores, but never (or at least, not yet) in the theropods.

In terms of locomotion, across the huge number of dinosaurs known and the variation in body plans from 1-kg bipedal theropods to >50-ton quadrupedal sauropods, there was a vast range of ability in terms of acceleration, top speed, and agility. Although dinosaurs were fundamentally terrestrial, many, if not all, could probably swim,[26] although few if any seem to have been even close to what might be termed semi-aquatic.[27] At least a few dinosaurs may have climbed trees,[28] and some could dig relatively well.[29] This is, of course, in addition to the flight of not just birds, but also various theropods close to the origins of birds that were at least gliders, and perhaps also included some capable of powered flight.[30]

Numerous dinosaurian taxa are known from mass mortality sites, suggesting that large numbers of individuals died together (see [31]), and there are also many extensive tracksites that show footprints of what are likely to be members of the same species moving and even foraging together (see [32]). Doubtless, many of these taxa were gregarious, or even social, and spent considerable amounts of time living in groups with the potential, at least, for complex social interactions. However, the evidence for sociality in dinosaurs has often been overstated;[33] almost any indication of two skeletons of a species found together has been used at times to infer complex behaviors in a species or even an entire clade.

There is strong evidence at least for sociosexual signals in dinosaurs,[34] so many species were signaling to each other (and to other species) even if they were not habitually living in groups. The most obvious of these signals include the many crests, frills, and horns of various lineages, but also include at least some of the feathers present on theropods, and display may have been an important component of early feather evolution.[34] The exact nature of any display behaviors is all but impossible to determine, though there is strong evidence of some large theropods engaging in ritualistic "scraping" courtships, in which, just as some modern birds do, pairs of animals would scrape the ground with their feet, leaving distinctive marks which have, rather incredibly, been preserved.[35]

Such interactions between dinosaurs would not always be simply about communication, as there is strong evidence of antagonist behavior in dinosaurs, especially between conspecific animals.[36, 37] Various preserved pathologies in fossil bones show that these animals fought one another, leaving serious injuries at times; numerous healed, and occasionally infected, bite marks (tyrannosaurs) and stab wounds (horned dinosaurs) are known. Although harder to diagnose, there is also some evidence for interspecific combat between dinosaurs, in particular for ornithischians fending off predatory theropods (see [38]). In addition to pathologies identified as resulting from combat, dinosaurs also inevitably accumulated injuries from daily living, and show evidence of a variety of deformities and pathological bones, including multiple different afflictions in a single individual.[39] Evidence of specific diseases or infections is harder to determine, though work has progressed in this area recently,[40] and in this light some modern diseases have been tentatively identified in dinosaur fossils[41] (figure 1.3).

As the dominant terrestrial animals for the majority of the Mesozoic, dinosaurs were key components of ancient ecosystems. They

FIGURE 1.3 Life reconstruction of *Tyrannosaurus* with infections in the jaw similar to those seen in modern carnivorous birds, following [41]. Artwork by Gabriel Ugueto.

would have had a profound effect on their local environments, and on the selection pressures on both other dinosaurs and other species. Exact interactions are effectively impossible to determine, but dinosaurs must have put severe browsing pressure on various plants and predation pressures on various animals (dinosaurian and otherwise), and they would have been host to gut bacterial floras, parasites, and so on. Similarly, there are many other aspects of dinosaur biology and behavior that we can infer from the fossil record only with considerable difficulty or have been able to ascertain for only a handful of specimens or species. Dinosaurs must have been competing with some other species for food and water, but which and to what extent is unknown. We know the colors of a handful of individuals (though there was doubtless intraspecific variation and at least in some species, dimorphism, with males and females differing in appearance). Issues like temperature tolerance, division of labor in parental care, water requirements, and so on are essentially unknown, and many might be unknowable (figure 1.4).

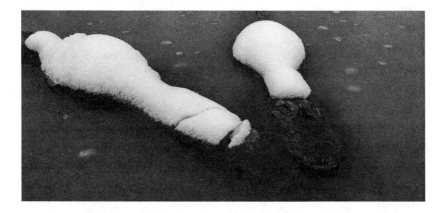

FIGURE 1.4 A relatively recent discovery is the tolerance for freezing temperatures by modern alligators (*Alligator mississippiensis*), suggesting a much greater range of physiology and behavioral options than previously recognized in the group. If such discoveries can be made in well-studied extant taxa, it naturally shows the difficulties of accurately reconstructing the physiology of ancient animals. Photograph courtesy of Jay Young and Jason McDonald.

Reproduction and Growth

Direct evidence of dinosaur sex has not been found, though clearly they must have mated. Although live birth is present in numerous extant reptile groups, no archosaurs are known to have given birth to live young (though it is suspected for one or two fossil clades), and so it is presumed that all dinosaurs laid eggs. Innumerable dinosaur eggs have been discovered, many in organized nests,[42] and even some entire nesting grounds are known. There is evidence for direct brooding by some of the most birdlike dinosaurs, including fossils of animals sitting directly on clutches of eggs.[43]

However, although post-hatching parental care is near-universal in extant archosaurs and was likely very common in dinosaurs, the extent of care, and the manner in which eggs and nests may have been incubated, and especially brooded, is uncertain.[42] The recent discovery of extremely long development times for some dinosaur embryos,[44] however, complicates this issue further, since it seems unlikely that large dinosaurs would have been capable of guarding nests for over three months at a time while still being able to forage and support themselves.

Given the limits of egg sizes, larger species would have grown through multiple orders of magnitude from hatchlings through to full-size adults, a pattern unusual among modern terrestrial vertebrates. In general, dinosaurs grew rapidly at a young age before growth slowed greatly as they reached larger sizes.[45] Unsurprisingly, dinosaurs did not start out as simply small-scale versions of adults, and changes during ontogeny, at least in some cases fairly dramatic,[46-48] would have resulted in their occupying multiple different ontogenetic niches as they grew.[49] Most notably, many structures that likely acted as socio-sexual signals were small in juveniles and only began to develop later in ontogeny, at or close to the onset of sexual maturity.[50] Growth in general was also non-uniform, and high plasticity in growth rates and great variation in size for a given age are seen in some;[51, 52] size is often not a good guide to the maturity of any single specimen.

A major issue for understanding the growth and development of dinosaurs is the general rarity of juvenile animals. Various biases work against the preservation and collection of small and juvenile fossils (they are harder for paleontologists to find, they are more likely to have been eaten by carnivores, and are more likely to decay), but even lineages of giants like sauropods are known from few small individuals. Compounding this problem is the difficulty in correctly identifying the age of a given specimen,[53] leading to conflicts over terms such as "adult," "subadult," "juvenile," and other age classes.[54] Combined with the differential growth trajectories and anatomical changes during ontogeny, this has led to considerable debate and/or confusion over the taxonomic identity of many juvenile dinosaurs (see [48]). Inevitably, therefore, examining the growth and development of dinosaurs is difficult when it is uncertain to which taxa various juvenile specimens may belong.

Brain and Senses

Most important, when considering dinosaur behavior, is their cranial capacity and the sensory input to their brain. Dinosaur brains are not preserved in the fossil record (bar one possible, and very notable, exception[55]). Data on dinosaur brain shape and structure is therefore derived from natural molds of the braincase or endocranium (the bony "inner skull" that encases the brain) or, more commonly now, digital models based on scans of dinosaurian skulls.[56] Since vertebrate brains have a largely stereotypical structure, major elements of the brain—most notably the olfactory bulb, cerebrum, optic lobe, and cerebellum—can be identified in dinosaurs, and give an approximate picture of their relative faculties. Paleontologists are therefore rapidly gaining an increasing understanding of dinosaurian senses.[57]

Dinosaurs lived in a complex world, and the variety of their ecologies inevitably led to a wide range of specializations in their capacity to sense and interact with their world. Dinosaurs were almost certainly tetrachromatic, and as with modern birds, would have been able to see into the UV spectrum.[58] This means that they would have seen a greater range of light frequencies than we do, and so a greater spectrum of colors. Work on the size of the olfactory bulb is also being used to determine the sense of smell in dinosaurs,[59] so it is becoming possible to begin to piece together the sensory systems of single species and the evolutionary history of these senses[60] based on the structures of their brains.

In addition to the components of the brain, the bony skull can also provide excellent evidence for the senses of dinosaurs. Some theropods, for example, had extraordinarily large eyes,[61] and this would have produced high visual acuity and/or the ability to see in low-light conditions. The shape and structure of the inner ears of some theropods have been elucidated and can be used to reconstruct the likely ranges of their hearing. Moreover, this reveals that some had asymmetric ears. This asymmetry allows animals to pinpoint the direction of sounds more effectively, and so would have been important for animals operating in low-light conditions,[62] thus giving further indications of the ecology of species. The presence of numerous complex foramina in the jaws of some large theropods has been used to argue for enhanced facial sensitivity in these species,[63] and although it is unclear quite how this would have been used or how sensitive it would have been, dinosaurs would have certainly had mechanoreceptors in their skin and so would have been receptive to touch.

FIGURE 1.5 Reconstructed brain of a *Triceratops* based on a 3D scan of the endocast. From this, major information about the size and structure of the brain, its component parts, and the inner ear can be determined. Figure courtesy of Ashley Morhardt.

Endocasts of sufficient quality to reconstruct the brain of dinosaurs in detail are rare, although some exceptional cases do exist, and while most are from larger, later species, this includes animals from the Triassic (see [64]) and tiny juveniles (see [65]) (figure 1.5). Studies of dinosaur brains have advanced significantly in recent years thanks to increasing attention being paid to these (see [66] for a recent review). Determining the intellectual capacity of even modern animals is difficult, and work on reptiles has been limited compared to that on birds and mammals, in turn limiting the scope for comparison with dinosaurs. Even so, it is possible to look at issues such as brain volume compared to body size and calculate the encephalization quotient (EQ) of dinosaurs for an estimate of their intelligence. (This can go awry, and the recent suggestion that *Tyrannosaurus* was comparable in intelligence to a chimpanzee has been subsequently squashed—see [67].) The EQ measure is a complex one and is not just a pure comparison of body size to brain volume, but scales with size—and different calculations are typically done for mammals, birds and reptiles.[68]

There are clearly major variations known in dinosaurs, and major uncertainties, too. For example, it has been noted that typical dinosaur braincases may have contained as little as 50 percent true brain tissues,[69] but Martin Brasier and colleagues[55] note that although iguanodontian dinosaurs had been given a reptile EQ range from 0.8–1.5, their specimen of an actual putative brain suggested that a much greater amount of brain tissue in an endocast could raise the EQ to as high as 5. For comparison, reptiles have an EQ from 0.4–2.4, extant crocodilians have an EQ of 0.9–1.1, and avians have a value that is typically 6 or above (though the range here is huge, running from 4 to over 28—all data from [68]), thus potentially putting some dinosaurs on a par with birds rather than reptiles.

Other values are of course known for various dinosaurs, calculated using various methods and differing brain-to-endocavity ratio (BEC) values. A good summary of this is given by David Evans,[70] who also provided some averaged estimates for various clades at the time. Among ornithischians, the hadrosaurs were at 2.8, horned dinosaurs 1.4, and armored dinosaurs 0.7. The Sauropodomorpha were 0.6, with various theropods ranging from 1.6 in allosaurs to 2.2 in tyrannosaurs and even as high as 7.1 and 8.6 for some of the most derived and birdlike clades. Some of these values come from only one or two studies of taxa, and some cover an enormous range of species and indeed an enormous range of values, but they are at least a good starting point. Collectively they suggest that, perhaps unsurprisingly, theropods (especially those closer to birds) were more intelligent than the sauropodomorphs and ornithischians. (See also [71] for a more recent list of dinosaurian taxa with EQ calculations.)

Uncertainty remains, though, as to how to calculate how much brain there is. Research by Daniel Jirak and Jiri Janacek[72] noted that the BEC ratio varies through ontogeny in crocodilians and could be as little as 29 percent in large adult tyrannosaurines, but a full 100 percent in the ancestral tyrannosaurs. In general, some of these lower values seem unlikely, and recent studies have pushed these values up and away from 29%–50% to more like 60%–70% and up (see [73]), so some of the older values given above that favored a 50 percent calculation are likely underestimates. In short, calculating an EQ or

BEC for any dinosaur is not easy. In addition to these varying values, quite how much an EQ tells you about the mental capacities of an animal is also questionable.[74] It is, though, probably safe to say that on average, dinosaurs were more intelligent than many extant reptiles, approximately as intelligent as living crocodilians and more intelligent squamates (lizards and snakes), but often less so than many modern birds.

For all that, brain size mostly simply correlates with body size, and this factor alone explains most variation in animal brain size. Thus any other extrapolations on intelligence immediately become problematic,[75, 76] while inconsistent considerations of variations in neuroanatomy and evolutionary trajectories can lead to misleading assumptions and mistaking noise for signals.[77] In short, dinosaur brains may be a leading clue to dinosaur behavior, but they can also be greatly misleading. The nature of dinosaur intelligence, therefore, and our ability to assay and even test hypotheses about aspects such as tool use (see [78]) are then inevitably challenging, to say the least.

Summary

It may sound trite to say that dinosaurs were real, living animals, but this point seems often overlooked because of the gaps in the fossil record that constrain researchers to focus on what we can work out from skeletons, footprints, and occasionally preserved soft tissues such as muscles and feathers, and far less on matters that are little known or unknown (figure 1.6). While it is difficult or even impossible to determine many aspects of dinosaur biology (or they are known from such limited data that broad inferences remain dubious), it should not be ignored that these were real animals that moved, mated, fed, excreted, competed, fought, died, and evolved. The gaps may be frustrating, but the rest of dinosaurian biology should not be forgotten, even if it is unknown. This brief overview should, though, provide a picture of the basics of dinosaur biology, and provide sufficient introduction to these animals to appreciate their diversity and disparity and to set the scheme for our basic knowledge of their lives.

FIGURE 1.6 Skull and life restoration of the Lower Cretaceous British spinosaurid *Baryonyx*. Although known from a well-preserved and relatively complete skull, exactly how it looked and behaved is uncertain, with numerous questions outstanding, as the evidence remains limited. Artwork by Gabriel Ugueto.

CHAPTER 2

Studying Dinosaur Behavior

Dinosaurs would have eaten, slept, bathed, groomed, courted, mated, built nests, laid eggs, walked, climbed, swum (and in some cases flown), called, defecated, regurgitated, learned, fought, played, migrated, hidden, and hunted at various times in their lives. But how many of these behaviors can even be captured in the fossil record? It would seem most unlikely for an animal to die while grooming itself and be preserved in a way that captured the act. Footprints might capture the moment two dinosaurs mated, but even so, could we correctly recognize it if we did find some odd complex interaction of two sets of tracks? These are the challenges of studying the behaviors of animals that have been extinct for over 65 million years, whose remains are dominated by skeletons that tend to be incomplete and that often were not buried till long after decay had set in.

The Fossil Record

In ordert too study and understand the behavior of dinosaurs, it is not just important to appreciate them as real and living animals, but also the nature of the available data from which paleontologists work. It is rather trite to note that dinosaurs are long dead, but this is also of superlative importance when trying to work out how they behaved when they were alive, and how the processes of death, decay, burial, fossilization, and discovery can influence or bias our perceptions of them and the data available.

At the most basic level, fossils can be split into *body fossils*, which are the actual remains of once-living animals, and *trace fossils*, which are the remnants of their activities[79] (figure 2.1). Thus bones, teeth,

FIGURE 2.1 Body fossils: above, bones of various dinosaurs from Dinosaur National Monument in Utah. Trace fossils: below, trackways left by small theropod dinosaurs in Zhucheng, China. Bonebed photograph courtesy of Rebecca Hunt-Foster, trackway photograph by the author.

armor, and preserved skin or feathers would be considered the former, and footprints, nests, and fossil feces (coprolites) the latter. There is, though, inevitably some overlap and odd intersections. For example, eggs are traditionally considered trace fossils, though eggs may contain embryos which clearly are body fossils (see [80]), preserved within nests that are trace fossils in their own right. Bite marks on bones left by carnivores are similarly both trace fossils (of the behavior that left the bite) and body fossils (the bone that was bitten, and potentially also teeth lost during the action of biting); and there are rare cases of skeletons found alongside tracks[81] or with nests of eggs.[43]

Body fossils of dinosaurs are dominated by bones and isolated teeth, as these are already highly biomineralized parts of the body, are not often consumed by carnivores and scavengers, and will be resistant to decay and thus survive long enough for a chance to be buried and preserved.[10] Other tough body parts such as ligaments, claws, and some skin structures (large scales, thick feathers) are rarely preserved compared to bones and teeth, but are much more common than tissues like muscles and viscera, which are both preferentially eaten by larger animals and decay rapidly from bacterial or similar action, and so are extraordinarily rare in the fossil record.

Footprints and related tracks (e.g., swim tracks, resting traces) are the most common trace fossils (see [82]), since animals are constantly moving, and in wet areas such as floodplains or riverbanks there is a good chance of local flooding that can preserve them. Nests are common in some regions, though normally dinosaurs would avoid areas that were likely to flood, as it would kill any developing embryos. Trace fossils are often difficult to attribute to species or even to broad clades, since the foot shapes and proportions of many taxa are similar, and traces can also be greatly distorted or present unusually, depending on the nature of the substrate. For example, dry sand will produce very different tracks than wet mud, even if the same animal was moving at the same speed across both; and many different groups have similar eggs, making them hard to assign to a given clade unless embryos are preserved inside.

The field of *taphonomy* essentially covers everything that happens to an organism from the moment it dies through to its discovery and excavation.[83] A fossil represents only a moment in time, and with body fossils, at least, it ultimately represents the moment that the animal was buried, and not usually the moment it died (though for a few unfortunates this was both). All manner of processes may have affected it both before and after burial that render it somewhat different from its original position and condition (figure 2.2). Although information like preserved stomach contents will represent a genuine record of its most recent activities before it died, it could have been exposed on the surface for weeks or even months before it was buried. When attempting to work out the lives of dinosaurs, it is therefore of critical importance to consider the geological context and taphonomic history that produced the specimens in question.[84] For example, a huge group of dinosaurs all found together might suggest that they were living (and died) together in a herd, but if there is strong evidence that the region was experiencing a major drought at the time, this may mean they were independently forced together at the last remaining water source, rather than having naturally moved together as a group. Similarly, shed teeth from a theropod alongside the skeleton of a herbivore might indicate that it was feeding on a kill, but shed teeth are very common and can wash into basins, so were other bones present, or is there evidence of a watercourse that might have brought tooth and bones together?

Even after issues such as alterations by the actions of living animals on the dead, other factors may alter specimens. Bodies can be washed down rivers or float out to sea, moving them from their original habitats.[85] Sun, rain, and cold can all break or alter bones even when they are buried, and geological processes can distort and twist them. There's a good reason so many dinosaurs are represented by incomplete skeletons and incomplete bones, and why so much work has been put into digital restorations and reconstructions of animals, in which mirroring elements or copying them from close relatives allows much more complete models to be built for analysis (see [86]).

Added to all of this are a series of biases in the fossil record that can distort and alter the perception of the reality of dinosaur biology if they are taken at face value.[87] For example, different ecosystems will preserve specimens to different degrees—bodies or tracks need to be buried, and there is little deposition in mountainous regions, but a lot on floodplains. Deserts are very good at preserving things since there are few scavengers and little decay, and so many more bodies will be buried, despite the often low animal populations. Contrast this with rainforests, which will support huge numbers of animals and may also

FIGURE 2.2 Organisms can undergo a number of alterations prior to fossilization and discovery, and so a fossil cannot usually be taken to be a true representation of the individual in life. Seen here is the body of a skua (*Stercorarius*) decaying on a beach, having lost most of its feathers and in an unnatural posture. Photograph by the author.

experience frequent flooding, but the decay rate is so high, coupled with acidic soils, that bodies will often be destroyed before they can be covered. We also have biases against juvenile animals, since their less-ossified bones preserve less well and they are more susceptible to being consumed whole by carnivores;[88] but also, smaller specimens are less likely to be preserved in general.[89] Large animals like sauropods are easy to find, but then they are so big they are rarely completely covered, and their very fragile skulls and pneumatic vertebrae often decay or are easily broken, while the robust limb bones survive. Similarly, transport of bones or whole bodies can break or damage bones too, and will also sort them—moving smaller bones further and separating out elements so that some information can be systematically lost or destroyed (see [90]). Teeth are exceptionally common, since not only did dinosaurs typically have many, but they would be shed and new ones grow constantly, so an animal might produce thousands or tens of thousands in its lifetime. With their strong enamel coating, teeth are also very resistant to being worn away, and so they tend to be preserved. Finally, there are biases in the field, not only in what paleontologists can find, but in what is then excavated and published on. New or rare species are more likely to be collected or described than well-known ones, and there is more focus on taxa that lead to the origins of birds, or that are unique in terms of features like large size.

The very fact that dinosaurs are fundamentally rare in the fossil record also brings in additional issues. Dinosaurs might be thought of as very common fossils, and compared to some groups they are, but the global distribution of dinosaur fossils is clearly wildly variable in terms of the numbers of species in different places and at different times in the Mesozoic (see [4]). We do, on average, have very few dinosaurs from the Middle Jurassic, for example, yet a handful of beds of exceptional preservation preserve thousands of specimens representing dozens of species.

Common behaviors may be difficult to capture (figure 2.3), while rare events may not be captured at all. When activities like mating may have taken place only for a few minutes a handful of times in the life of an animal, this is unlikely to be captured in the fossil record for a given species or even for the entire clade, so numerous key behaviors may go entirely unrecorded. The other side of this is that because there are so many fossils overall, even very rare events will be captured on occasion in one specimen, and it is too easy to infer that this is normal. For example, a small carnivore might have been predominantly insectivorous and only rarely consumed vertebrate

FIGURE 2.3 Dinosaurs must have spent portions of their lives asleep or at rest, but it is very unlikely that they could be preserved while retaining such a posture. Some exceptions are known, however, such as the juvenile troodontid *Mei* shown here, in a resting posture strikingly similar to that of modern birds such as this mallard duck (*Anas platyrhynchos*). Note that the tail of the dinosaur is broken but was originally wrapped around the body. Photographs by the author.

prey, but bones have a much higher preservation potential, and so stomach contents would preserve these and not the remains of invertebrates. This rarity issue is true of specimens and fossils in general, since often we are working from only one or a handful of individuals that may themselves not be representative of a population or species.

Thus, although pathologies and genetic diseases often come with clear indicators that can be determined from the anatomy of an animal, others do not and so can give a false impression of the frequency of afflictions of dinosaurs. Similarly, it is at least possible that all specimens in a sample might be male members of a highly dimorphic species and would thus present a rather anomalous impression of the species as a whole.

Specimens are, of course, single data points. Although some mass mortality sites might capture hundreds or even thousands of individuals,[91] typically a fossil specimen represents a single individual. It might be an adult or a juvenile, male or female, buried in summer or winter, or be atypical of its species, and we often would not be able to pick between any of these alternatives. As a single animal, it gives us no idea of the degree of intraspecific variation, changes over ontogeny (growth), or change over time for the species, what seasonal or regional variations there were in the population, or how widespread it was (figure 2.4).

FIGURE 2.4 Species that are typically solitary can aggregate under environmental stresses, seasonal changes, or other events. Stressed situations can increase the chance of mortality and so produce fossils that suggest animals lived together when they did not. Here, three African rock pythons (*Python sebae*) emerge from a shelter at the start of a rainy season, but this is not a gregarious or social species. Photograph by the author.

All these issues add up to potentially mislead researchers about the true nature of the fossils they are studying. While we have a good understanding of many of these pitfalls and the ways they can confound data, analyses, or conclusions, there is only so much that can be done on any given specimen and inferred for a species. This context must be considered when examining and interpreting the fossil record and trying to draw possible behavioral data from it. This is especially important to keep in mind given what we know in extant animals about the intraspecific variation in behaviors and the plasticity of the behavior of individuals. Even animals widely regarded as purely solitary and actively antagonistic to conspsecifics may nevertheless form aggregations at times, and social behavior in particular can vary wildly between juveniles and adults, males and females, and by season or situation (e.g., as seen in cheetahs, *Acinonyx jubatus*[92]).

Animal Behavior

Although dinosaurs have a much longer history of formal scientific study than does animal behavior, the latter is a much larger field as a whole (at least in terms of researchers and publications). When it comes to interpretation and understanding, the real foundations of behavior are perhaps still best represented by Niko Tinbergen's[93] four kinds of explanation for behavior—function, causation, development, and evolutionary history.

Function relates to how a behavior might affect the ability of an animal to survive and/or reproduce (i.e., to increase its evolutionary fitness). *Causation* relates to the stimuli that cause the response: what information is the animal receiving, and what has it learned related to this, to produce this behavior? *Development* relates to changes during ontogeny and the differences in behavior that can occur with age and experience. Finally, *evolutionary history* refers to the relationship of behaviors to phylogeny and how those behaviors relate to the behavior of relatives and ancestors. In each case, a supportive question should not simply be, "Why does the animal conduct this behavior?" but also, "Why not another one?"

FIGURE 2.5 The behavior of an animal being studied in the lab under controlled conditions can be repeated for multiple samples, something impossible for extinct dinosaurs. Here a Hermann's tortoise (*Testudo hermanni*) is being trained to recognize certain shapes associated with a reward. Photograph courtesy of Elisabetta Versace.

Clearly, not all of these can be easily studied or even inferred for dinosaurs (figure 2.5). We can expect that they were able to learn, and to receive and respond to stimuli appropriately, but what those stimuli were, and how they elicited and affected responses, is effectively unknowable. It is reasonable to infer that prey animals considered large theropod carnivores a threat to be avoided, but quite *how* they were perceived as such, the exact *response* to that, and how individuals *learned* that theropods were a threat, cannot be determined.

Importantly, the methods normally applied to the study of behavior in living animals are all but inapplicable to dinosaurs. The four main methods of studying behavior are through experiments, observation, comparison, and theoretical models.[94] Obviously, extinct animals cannot be observed in the wild or in captivity. We cannot experimentally change conditions to see how they respond, or observe a single individual across a lifetime or under different stimuli, and we often have only single data points to work from. Furthermore, there are limits to how much information we can draw from living animals: after all, our understanding of *their* behavior is often far from perfect. In any case, the behavior of extant animals is often not a good fit for extinct ones—although we can make educated guesses and estimates, it may not be appropriate to scale up our understanding of behavioral and ecological patterns and rules from large living terrestrial carnivores to giant theropods that were bipeds ten times larger than big cats or crocodiles.

Behavior in the Fossil Record

How, then, can dinosaur behavior be studied, given the problems with the fossil record and our inability to observe dinosaurs directly?

Although numerous avenues of data are closed, plenty remain. However, their application and interpretation in terms of behavior has been somewhat haphazard, and there has been a lack of rigor in the approach to assessing the available information. While testing of hypotheses can remain impossible, and much in the following chapters will rely to a degree on inferences and comparisons, extracting behavioral data from the fossil record requires a more systematic approach.

One major aspect of this is the sheer plasticity of behavior in extant animals and the occasional problems of matching morphology to activity. While lions (*Panthera leo*) mostly live and hunt in groups, most other cats do not, and some lions are solitary or switch between the two.[95] So evidence for a behavior in a species does not mean it was universal for that species (or even for that individual), and evidence for behavior in one species cannot necessarily be applied to even close relatives (figure 2.6). Morphology is often a superb clue to the general habits of species, but the diet of crocodiles can include fruit and other vegetation[96] despite their clear adaptations to a carnivorous, and even a specialist piscivorous, diet. Ultimately, therefore, the issue becomes one of what is known and what is likely, what is plausible, and what is possible.

Most behaviors are at least *possible*: almost any behavior could conceivably have been carried out by any given dinosaur at some point. More important, then, is what is plausible: that is, what are the kinds

FIGURE 2.6 The proboscis monkey (*Nasalis larvatus*) is unusual in spending time in and around water, but this behavior cannot be extrapolated to other primates, even close relatives. Photograph courtesy of Robert Knell.

of behaviors we might expect given what we know of the animal? Species with sharp teeth are typically carnivores, but it's plausible the animal ate a reasonable amount of plant matter as part of its diet. Closer in, we can say what is likely: a large carnivore likely tackled larger prey than a small species did. Finally, we might say that something is *known* about the behavior of an animal: it may be only a single data point, but the presence of, say, seeds in the stomach of a dinosaur shows that at least once, that animal was eating them. Confusion of these points, and incorrect or inappropriate extrapolation from one case to another on too little data, is a common issue.

A series of recurrent problems with work on the behavior of fossil animals were laid out by myself and Chris Faulkes in a 2014 paper[97] that covers some of the issues with the data itself, as well as how it has been (and continues to be) misrepresented in the literature. First, false dichotomies may be presented: for example, evidence for active predation by a carnivore does not rule out scavenging, or vice versa. Second, an incorrect claim about a behavior of an extant animal or clade may be used as a model for comparison to extinct ones.

Similarly, the evidence available might actually contradict, in full or in part, the proposed hypothesis for a behavior in a fossil animal. Next, there may be over-extrapolation from data or conflation of data, such as inferring that all members of a species or even a clade lived in groups based on one example of individuals found together, which themselves might not even have spent most of their time in groups. Finally, there is a problem with a lack of specificity and overuse of generalities, which make it hard to ascertain what is actually being proposed or intended: for example, saying that an animal engaged in "social behaviors" which is vague and could fit very different definitions of "social," depending on the context. The use of such undefined terms, or a failure to specify exactly what is being suggested or hypothesized, continues to plague studies and makes it difficult to put new data into context. In Hone and Faulkes[97] we proposed a set of guidelines for establishing and assessing hypotheses of fossil-animal behaviors that can help bring both clarity and rigor to the field:

1. Make it clear that a specific hypothesis is being established.
2. Be specific about terms used (defining them as necessary) or the extent of a behavior being assessed (e.g., "social" may cover a huge range of behavior[98]).
3. Is the hypothesized behavior considered to be obligate, common, or merely possible? (Animals that *can* climb are very different in their normal behavior from those that are arboreal.)
4. As far as possible, the behavior should be known to be present in an extant taxon.
5. Extant examples (and even extinct ones in some cases—see [99]) should be close to the taxon in question, either phylogenetically or in terms of a functional or ecological analogue.
6. Behavior is extremely plastic, so extant analogues should take into account known variation and causes of different expressions of behavior (such as seasonality; see [100]).
7. Functional morphology and comparative anatomy can provide a strong evidence base for behaviors.
8. Traits may be multifunctional, and how a trait functions now may be very different from its evolutionary origins and drivers

(e.g., elephant trunks are used in gathering food, drinking, communication, and even combat in addition to olfaction[101]).

9. Strongly selected traits need not be under habitual use (e.g., many animals might mate only once in their lives, but selection for anything that supports reproductive success would be strong).

10. Taphonomy and sampling biases must be taken into account, and consistency across multiple specimens should be sought.

11. Additional lines of evidence that are as independent as possible should be used to support a given hypothesis.

Some of these issues are easier to address than others. It should not be difficult to provide clear definitions of terms being used or formal hypotheses. On the other hand, many dinosaur species are represented by single fossils, or they come from fossil sites with a complex history that are difficult to interpret (see [102]) and limit what can be done to account for issues such as taphonomic history. As an example, it may be hypothesized that a newly discovered large dinosaur was herbivorous and that its adults predominantly ate leaves. This behavior is seen in numerous extant animals, so there are solid analogues for this in reptiles, birds, and large-bodied terrestrial mammals to form a specific hypothesis. Predictions can be made about the biology of this animal, based on the hypothesis, that would separate it out from others, and then multiple lines of evidence can be sought to support (or ideally, test) this. So: tooth shape can be compared to other leaf eaters; the localities from which the fossils were recovered can be checked for evidence of leafy plants; their skull muscles can be reconstructed to see how they would have bitten and whether they had weak bites; isotope data from bones and teeth can be analyzed to see if they ate plants; tooth enamel could be examined for microwear scratches to see if this matches other leaf-eating animals; and multiple specimens could be assessed from multiple sites to see if these are all consistent patterns. Other possible aspects of their behavior could be considered, such as whether they may have fought one another with their teeth; this might add some counterselection and could make the teeth less optimized for leaf eating, despite a specific diet. A hypothesized diet of leaves is not an especially complex behavior to infer or test in a dinosaur (the list of analyses suggested above is excessive), but even here there are multiple possible explanations for what might drive tooth and skull shape, and consideration of these is necessary to build up an accurate picture of what the animals likely could and could not do—and of what might lead us astray in our results.

Analogues and Extrapolation

Extrapolations from extant animals are a particularly important part of assessing dinosaur behaviors. Not only do living species give us a range of behaviors as a starting point for forming hypotheses about dinosaur habits, they can also place limitations or allow us to make major inferences about the behaviors of dinosaurs.

The most important concept is that of the *extant phylogenetic bracket* (EPB, see [103]). In its simplest form this allows for the inference of a behavior in dinosaurs based on their nearest living relatives (i.e., the dinosaurs are bracketed by extant relations), which are the birds (the direct descendants of dinosaurs and so literally dinosaurs) and the crocodilians. If a characteristic is present in both, then the default hypothesis would be its presence in dinosaurs as well. If it was present in only the crocodilians, then it was likely present in the ancestors of dinosaurs but was lost somewhere on the route to birds, and so may or may not have been lost at any point between the two. Similarly, anything present in birds and not in crocodilians could have been present in dinosaurs ancestrally or might only have appeared on the branch including modern avians.

This provides a powerful starting point, though with some obvious caveats. Birds are highly derived animals that ancestrally could fly and are limited in body size, while modern crocodilians have reverted to a semi-aquatic lifestyle and are all carnivorous. Both, therefore, have constraints and life histories that could have profoundly modified or limited their behaviors, and this must be borne in mind, but the EPB remains a robust starting point. If a behavior is exhibited by neither clade, then a very strong case would need to be made for it to be present in a dinosaur. Advanced studies of both bird[99] and crocodile[104] behaviors are now normal for studying dinosaur ethology, and it should be remembered that the "lesser" reptiles like lizards are known to engage in complex behaviors (see [105]) that are often incorrectly considered to be restricted to birds and mammals.

Looking at these animals also allows limitations to be placed on the dinosaurs in a similar, if less formal, manner. For example, some birds are capable of creatively solving difficult problems (e.g., crows, parrots), but these are real rarities among birds, and the earliest-branching groups like the ratites are far from this level. From this it is reasonable to infer that while the dinosaurs might have had the capacity to evolve very clever species, we would not expect this to be common, and dinosaurs generally are perhaps likely to have been closer in intelligence to the earlier-branching members of the avian clade. Similarly, if a behavior or pattern is present in animals such as lizards, then it would be unreasonable to assume *a priori* that this would be beyond the capacity of dinosaurs.

Other patterns are common across numerous clades or are a direct result of essential functional morphology that can be used to make reasonable inferences. Mechanical analyses are perhaps the most powerful of these, and so, for example, where it has been shown that the specialized narrow middle metatarsal of the feet of tyrannosaurs (the arctometatarsalian condition) greatly enhances locomotor efficiency,[106] then the convergent development of this feature in several other derived theropods immediately suggests a selective pressure toward a common solution, and that these animals also were efficient long-distance travelers (see [107]). Such biomechanical analyses are also important for ruling out competing hypotheses. For example, the large frills on the skulls of ceratopsians have been hypothesized to have had numerous different functions, including as a shield against predators and as articulation points for giant jaw muscles. However, these can be shown not to work biomechanically: no attachment points for muscles have been observed, and contemporary theropods were capable of biting through much thicker bones than the frills, thus effectively eliminating these ideas from consideration (see [34] for a review).

PLATE A A pair of the Early Jurassic theropod *Sinosaurus* engage in a hypothesized bonding ritual with mutual face rubbing. Such behaviors might well have been common in social animals or mated pairs. Artwork by Gabriel Ugueto.

PLATE B A standoff between the Late Jurassic stegosaur *Tuojiangosaurus* and the large allosauroid *Yangchuanosaurus* in central China. Although numerous depictions show fights to the death between large predators and potential prey, adult animals, especially those with defensive spines, would not be the preferred target for carnivores. Artwork by Gabriel Ugueto.

PLATE C The giant North African *Spinosaurus* in the act of fishing. While the ecology and behavior of this unusual animal is controversial, all agree that fish was a major part of the diet of this theropod. Artwork by Gabriel Ugueto.

PLATE D A group of the tiny alvarezsaur *Linhenykus* foraging for small invertebrates in the deserts of northern China during the Late Cretaceous. The ceratopsian *Protoceratops* passes in the background. Artwork by Gabriel Ugueto.

PLATE E A mixed group of large sauropods, *Apatosaurus* with a single *Brachiosaurus*, rest at night in the Late Jurassic of North America. In the foreground, the theropod *Ceratosaurus* walks by, uninterested in such large and dangerous prey. Artwork by Gabriel Ugueto.

PLATE F A pair of Late Jurassic *Allosaurus* in a standoff over the carcass of a young *Camptosaurus*. Combat between members of a species was likely preceded by posturing and stereotyped displays as antagonists sized up the potential risk of a conflict before committing. Artwork by Gabriel Ugueto.

PLATE G A group of the small Late Triassic theropod *Coelophysis* stop in a glade and engage in some of the huge variety of activities of which dinosaurs were capable, from foraging to grooming to defecating and resting. Artwork by Gabriel Ugueto.

PLATE H The British thyreophoran *Polacanthus* seen in a hypothesized mating posture. The wide pelvis and rigid armor of these animals would have made it difficult for them to copulate, and one solution to this issue would be a large phallus in the male. Artwork by Gabriel Ugueto.

PLATE I A pair of the huge North American ceratopsian *Triceratops* engage in a duel. There is good evidence that these animals fought head-to-head, leading to the formation of serious scars on their faces. Artwork by Gabriel Ugueto.

PLATE J The theropod *Tyrannosaurus* cannibalizes the corpse of another giant who has not survived the cold. A number of tyrannosaur fossils show feeding traces of conspecifics, and although cannibalism was probably generally rare, it clearly occurred. A number of small dromaeosaurs are also taking advantage of the available meat. Artwork by Gabriel Ugueto.

PLATE K A pair of the large abelisaurid *Majungasaurus* sleep through the day in the end Cretaceous of Madagascar. In the foreground, the small herbivorous crocodile relative *Simosuchus* walks past. Sleep would have been a major activity for large theropods, but it is hard to predict the duration and what stimuli, like this possible prey, would have changed their behavior. Artwork by Gabriel Ugueto.

PLATE L Foraging would have been a major part of the life of carnivorous animals. Here the small compsognathid *Sciurumimus* patrols the beach of the Late Jurassic of Germany looking for anything that might have washed up after a recent storm, but the body of a huge marine reptile, *Dakosaurus*, is not of interest. Artwork by Gabriel Ugueto.

PLATE M The huge carcharodontosaur *Meraxes* pursues a young rebbachisaurid (one of the diplodocoids) sauropod in the Late Cretaceous of Argentina. Young dinosaurs that were naïve and lacked the size or defenses of adults would have been the primary target of predators. Artwork by Gabriel Ugueto.

FIGURE 2.7 The mouth shape, body size, and presence of horns are all linked to the social ecology of antelope. Seen here, left to right, are the small and narrow-mouthed duiker (*Sylvicapra* sp.) that lives in dense bush; the large but narrow-mouthed kudu (*Tragelaphus strepsiceros*, here a male—females lack horns); and the large and wide-mouthed wildebeest (*Connochaetes taurinus*) from open plains. Artwork by Gabriel Ugueto.

More intriguingly, some aspects of functional morphology can be tied to other aspects of behavior, as seen for example in modern ungulates:[108] the size of the mouth (broad vs. narrow) is fundamentally linked to the size of the species and the foods that they prefer (figure 2.7). Smaller animals can support themselves on small and nutritious buds, new leaves, fruit, and the like, and so have small mouths that can selectively pick up choice foodstuffs. Large animals generally need to be less choosy, however; they have broader mouths to acquire fodder in bulk, and their larger sizes allow longer digestive time to break down coarse grasses and tougher leaves. These differences then feed directly into habitat occupation: small ungulate species can live in dense areas where select foods can be reached and cover is provided, but large species cannot forage in confined areas. This also then links to the social biology of these animals. Small species only need to forage in a small area, and so this can be defended, leading to animals often living in pairs and both sexes having horns to defend the territory from rivals. But large species need to range to find food, leading to their forming herds in which vigilance can provide protection, and promoting sexual dimorphism: males have large horns to display and defend a harem of females from rivals, but females lack them. Thus ultimately mouth size and body size can correlate with habitat type, feeding behaviors, social interactions, and degrees of dimorphism.[108] It is a complex arrangement, but one that could be near-inevitable, since the size required to digest tough food vs. the availability of cover or accessibility of small, nutritious food surely applied in the Mesozoic, and so similar arguments have been made for dinosaur behaviors based on this.[109]

Even allowing for these patterns and techniques available to assess dinosaur ethology, determining patterns of behavior, even some very basic ones, may be much more difficult than appreciated, and so remain unknown for most dinosaurs (see Case Study), or are based solely on inferences from studies of other relatives. Even so, some clear inferences can often be made through simple extrapolations, such as: feathered theropods would be well insulated and so were likely to be more capable of dealing with colder environments or conditions, and so would have had a broader range of activity levels than non-insulated animals of comparable size.

Moving Forward

Despite the negativity expressed through much of this section, a strong case can be made that paleontologists are getting an ever better handle on dinosaur behavior. The combination of new fossil finds (such as the discovery of filamentous coverings on ornithischian dinosaurs[110]), new techniques in analyzing them (e.g., CT scans of skulls to reveal brain shapes[57]), an ever better understanding of the drivers and correlates of behavior in living animals (including social behaviors[109]), and the sheer weight of studies (especially on bird origins[17]) means that the field is expanding rapidly and becoming more rigorous and certain in its outcomes. In many cases, the results might be that we have in the past overinterpreted data and are less confident now (see [111] on sexual dimorphism), but I would personally always favor a degree of uncertainty or lack of confidence in a result, rather than embracing an interpretation confidently that may be wrong and will mar the approach of future studies with a misleading starting point.

As outlined above, numerous lines of evidence are now available to researchers to help uncover dinosaur behavior. In particular, the development of more studies on the mechanics of teeth (and their associated musculature) and bones have proven a revelation in terms of determining how animals could (or could not) move. These can provide an incredible basis for forming new hypotheses or testing existing ones (figure 2.8).

Pitfalls remain, however. Although mechanical studies can be an excellent way to rule in or out certain activities, such works do not usually provide absolute evidence for how a structure was used. An analysis might show that the jaws and teeth were strong enough to bite into bone, but this does not necessarily mean that the animal did this, or that this was the primary driver for the evolution of strong jaws and

FIGURE 2.8 Life reconstruction of the bizarre little gliding theropod *Yi* from China, shown engaging in a hypothesized grooming behavior on its feathers using its large front teeth. Such a behavior could be tested for by looking for microscopic grooming grooves on the teeth such as those seen in some mammals.[99] Artwork by Gabriel Ugueto.

teeth. Similarly, the sheer plasticity of the exhibition of behaviors of living animals (by single individuals, within species and clades) means that extrapolations can be problematic (especially from single specimens, which are the norm for dinosaur fossils). Social behaviors in particular can vary very markedly, and so calling one species social because another closely related species shows some fossil aggregations is likely to be erroneous. Hence we return to the importance of multiple lines of evidence that can mutually support single conclusions, thus allowing specific claims to be tested and verified, though generalizing from the results may be difficult. This process does, though, at least provide a solid base for determining dinosaur behavior.

CASE STUDY: **Behavior in *Tyrannosaurus***

Tyrannosaurus rex has become a model organism for dinosaur research,[112] and despite the limited number of good skeletons that have been recovered,[48] probably more scientific papers have been written about it than any other dinosaur. Even so, huge gaps exist in even our most basic knowledge and understanding of this animal; considerable uncertainties remain, and alternative interpretations are possible from the available data. (Many of the points raised below are covered in more detail in later chapters; the notes here are kept deliberately brief.)

Tyrannosaurus has been found in multiple different US states and in Canada,[4] and given its range, it must have occupied a number of different environments and climates, but we do not know which, or if any were favored, and the true extent of its distribution is unknown.

It was certainly carnivorous:[23] we have evidence of (attempted) predation,[113] and we can infer scavenging from other large tyrannosaurs,[114] but not in what proportions, and prey preferences are unknown, though *Tyrannosaurus* had an extremely powerful bite capable of breaking into strong bones.[115]

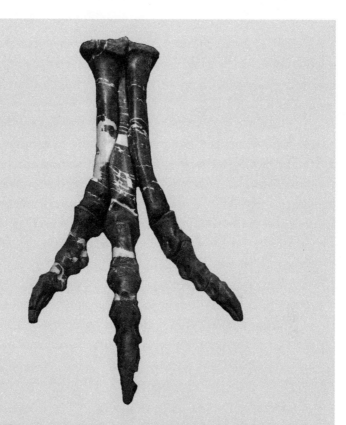

FIGURE 2.9 The arctometatarsalian condition, seen in the foot of a *Tyrannosaurus*. The pinched middle metatarsal seen at the top of the foot is an adaptation for efficiency, suggesting that animals with this foot structure often traveled long distances. Specimen is from the US Bureau of Land Management. Photograph courtesy of Eric Metz and the Museum of the Rockies, Montana State University.

We have a good understanding of its locomotor abilities and data on its ability to accelerate and turn (see [116]), and the structure of the foot suggests it was an efficient long-distance walker (figure 2.9). We know that it had excellent binocular vision and large eyes,[61] and good senses overall, but the suggestion that it may have been crepuscular or nocturnal[117] has yet to be tested.

Tyrannosaurus has been proposed to have been sexually dimorphic (see [118]), but this has not been demonstrated under rigorous testing,[119] and it remains uncertain whether some limited dimorphism was present. Similarly, there are some good reasons to think that the small horns over the eyes of this animal were used as sociosexually selected displays[34] based on analogy with many other animals, but this has not been assessed formally.

There are suggestions of other tyrannosaurs[120] engaging in social behaviors, but there is nothing to directly support this in *Tyrannosaurus* specifically. We also know that, like other tyrannosaurus, it engaged in intraspecific combat,[121] and while injuries to the face from biting start to appear as the animals enter their reproductive years,[37] we do not know if these fights were inter- or intrasexual, or linked to territorial control or competition over other resources (figure 2.10).

There are no known eggs, nests, or embryos for *Tyranno-saurus*. We do not have even indirect knowledge or evidence of courtship and mating behavior, numbers of eggs, pre- or post-hatching parental care, or other aspects of reproductive

FIGURE 2.10 Life reconstruction of the head of an adult *Tyrannosaurus* showing numerous scars from intraspecific combat. Artwork by Gabriel Ugueto.

behavior. Juvenile specimens are rare, although we do have a good understanding of their growth and ontogenetic development,[48] and some sense of population structures.[122] *Tyrannosaurus* juveniles were quite different-looking from adults, and not simply small versions of them; they would have occupied ecospace differently, hunting different prey and probably in a different manner,[49] but exactly what they were doing, and when in their growth cycle, is not known.

Finally, it is worth making a brief comment on the famously small arms of *Tyrannosaurus*. Despite their small absolute size and their progressive reduction across the tyrannosaur lineage, huge numbers of ideas about possible functions of the arms have been published, including many that are poorly founded, inadequately thought through, or untested. It's been advocated that the arms were reduced in size to avoid injuries during speculated feeding frenzies;[123] to help push over potential prey;[124] to hold prey while attacking them,[125] and many more, though arguably the strongest explanation is still that they were greatly reduced compared to the arms of other predatory theropods simply because they were little used.

In short, we do know a lot about the biology, ecology, and behavior of *Tyrannosaurus*—perhaps far more than many nonspecialists might realize—but there are also huge gaps in our most basic knowledge and understanding of behaviors that would be taken for granted in studies of living animals: Were they gregarious or asocial? Did both parents contribute to raising offspring? What are their preferred prey species and habitats? And to reemphasize the point, this is regarding one of the best known and most studied of dinosaurs. This should then give an idea of the limitations of data with which we are working when it comes to the study of the behavior of dinosaurs. Therefore, of necessity we must infer, interpret, extrapolate, and even make best guesses as to what dinosaurs were doing and why.

CHAPTER 3

The Basis of Dinosaur Behaviors

Our understanding of dinosaurs has come a long way in the last 200 years. The scientific conception of these animals has moved from lumbering, sprawling evolutionary failures to one of the most successful groups of tetrapods, with a diversity of species and disparity of forms to rival most clades (figure 3.1). But they were, above all, living animals. So while the focus in media representations will always be on thrilling fights to the death between a *Tyrannosaurus* and a *Triceratops*, the daily lives (and even annual lives) of dinosaurs would have been more mundane. Later chapters will cover major areas of dinosaur behavior that have been researched extensively; the focus here is on "basic" biology—fundamentals of their behavioral ecology such as movement, posture, and physiology—that would have set the framework for the daily lives of dinosaurs and determined or limited their behaviors.

Activity Patterns

It is difficult to establish the general activity patterns of dinosaurs, though clearly their sensory apparatus can be a strong indicator. Most obviously, nocturnal animals have absolutely (not just proportionally) large orbits to accommodate large eyeballs, providing good visual acuity in low light (see [126]). Attempts have been made to identify possible dinosaurian candidates for low-light activity (see [127]) based on orbit size and the size of the sclerotic ring that supports the eyeball, but with limited success. Large eyes may also, though, correlate simply with high visual acuity at long distance, and it is notable that the largest eyes seen are in giant tyrannosaurs who also had the best

FIGURE 3.1 Life reconstructions of the dinosaurs *Iguanodon* (top) and *Megalosaurus* (bottom), based on the famous Crystal Palace models in London that were built in the 1850s. They were considered at the forefront of paleontology at the time. Artwork by Gabriel Ugueto.

binocular vision,[61] indicating predatory animals that were reliant on vision in any light (figure 3.2). Taking a different approach, Scott Hartman and colleagues[128] inferred likely nocturnal, crepuscular (dawn and dusk), or cathemeral (any time of day or night) activity for the small Late Triassic theropod *Coelophysis* based on its modeled thermal physiology and the prevailing climate of its environment.

Data on eye size and/or other senses therefore can be integrated with other information to build up a putative pattern of behavior. The small insectivorous alvarezsaurs, for example, have proportionally large eyes in their heads (though the absolute size is small), which may indicate nocturnal habits, but more notably, they have asymmetric ears similar to those of owls, suggesting exceptional hearing that can accurately determine the spatial location of sounds,[62] providing strong corroborating evidence that these animals operated in low-light conditions.

FIGURE 3.2 Sclerotic rings in the eyes provide a measure of the size of the eyeball and the visual acuity of the animal. Shown is the skull of the theropod *Sciurumimus*, showing that the ring sat inside the orbit. Photograph by the author, used with permission of Raimund Albersdörfer.

The inner ear can be scanned in well-preserved dinosaur skulls, enabling determination of the size and shape of auditory canals and thus indicating the kinds of sounds (high or low pitch) to which animals might be attuned,[57] potentially providing a proxy for behavior.[129] It was shown by Stig Walsh and colleagues[129] that across multiple reptile and bird taxa, the size and shape of the endosseous cochlear duct showed positive correlations with diverse features of biology, including "measures of hearing sensitivity, vocal complexity, sociality and environmental preference," strongly suggesting that ear canal shape can be used to make inferences about behavior. That said, this is a good exemplar of one common problem with dinosaurs, namely their huge size compared to extant taxa. Extending trend lines from birds (largest extant examples ca. 140 kg) and modern crocodilians (1 t) to the largest theropods (7 t) and sauropods (>70 t) is clearly potentially problematic.

Olfaction has also been assessed in various taxa, most notably theropods, using olfactory bulb size in the brain as a proxy for assessing the sense of smell. Surprisingly, this showed an increase in olfactory capacity in early birds, and that many theropods were less specialized in this regard, although olfactory bulbs were generally larger in predatory than herbivorous lineages,[59] with tyrannosaurs having especially large bulbs.[67] Some ornithischians have also been assessed, and, for example, the small olfactory bulb was used to infer a weak sense of smell in the large neoceratopsian *Triceratops*,[60] suggesting less reliance on this feature.

Dinosaurs would have had other senses, and at least some work has been done on their mechanoreception. Foramina on the jaws of large tyrannosaurs have been used to infer the passage of numerous facial nerves, and by extension, a sensitive jaw that could have functioned not only in foraging, but even in aspects of behavior such as courtship.[130] Such features are not unique to tyrannosaurs, however, and it has been noted that this pattern is common in theropods and may relate more to a normal distribution of nerves in dinosaurs than to any especially tactile behaviors or specializations (see [63]). Drawing data from a different source, Phil Bell and Christophe Hendrickx[131] compared the scale structures on the tail of the well-preserved small (?compsognathid) theropod *Juravenator* to similarly structured sensitive scales in extant squamous-skinned tetrapods, and suggested that these scales functioned in mechanoreception in *Juravenator*, though their possible functional role is uncertain.

As with many areas of dinosaur biology, data is limited here to a relatively low number of studied taxa, and the results, if not indeterminate, generally provide only a degree of likelihood rather than any firm answers about activity levels and sensory data. Even so, the acquisition of data on the sensory apparatus of dinosaurs can provide a lead as to their ecology and behavior, and this has the potential to integrate well with other lines of evidence. For example, among African herbivores, larger animals forage and move less at night and avoid cover to reduce risk of predation by larger carnivores; however, smaller species may be more active at night when they are better concealed, especially from predators such as eagles that are active only in the day.[132] While not considered here, seasonality should not be ignored, as it can have profound effects on behaviors, most obviously those associated with breeding, feeding, and migration (see [100] on birds).

Habitat Occupation

Specimens of dinosaurs are only found where they are buried, and this may not represent where they lived normally. As noted earlier (under Studying Dinosaur Behavior), biases of preservation mean that taxa may be locally abundant but rarely preserved, and vice versa; and transport between environments is also possible.[133] For example, ankylosaurians of various forms tend to be more common in marine sediments, and it has been suggested that that these animals preferred coastal environments or perhaps estuarine channels. However, a detailed study has shown that these patterns are largely the result of local extinction and the absence of various clades in certain regions. It was was found that some of the nodosaurids may indeed have had affinities for coastal regions, though the pattern is complex.[134] On top of that, ankylosaurid and nodosaurid corpses appear to have floated differently, which may have led to differential transport potential, and so any animals that were washed into channels may have traveled very different distances, and this could disrupt attempts to trace such patterns of habitat occupation.[85] At the opposite end of the spectrum, we have clear evidence of desert-dwelling animals in areas largely devoid of large water bodies.[20] Based on their distribution, dinosaurs clearly covered a vast range of habitats.

Inevitably, larger species of dinosaurs would have ranged across larger areas and would therefore typically be found in more varied environments than small species that could specialize to one habitat or climate (or even microclimate). Fossil excavations are limited by which fossil beds are accessible, and these are generally relatively small in area, limiting our ability to determine how much space a given species might have covered. There are exceptions, though: for example, the Late Jurassic Morrison Formation that covers much of the western part of the United States has preserved specimens of the same sauropod species more than 1,000 km apart.[133] On the other hand, we also (predictably) see distinct faunas occupying distinct habitats. Two Late Cretaceous formations from North America demonstrate this well: there is little overlap of taxa between the faunas of the dry and inland (and perhaps also predictably less diverse) Two Medicine

Formation and the fluvial/floodplain Judith River Formation. These two formations are contemporaneous and close to one another,[133, 135] and so their distinctive faunas point to localized habitat occupation and adaptation by the various species.

Animals may not always occupy a single location throughout their lives, or even throughout the year, and it has long been suggested that many dinosaurs migrated. This is based primarily on mass mortality sites representing hundreds, or even thousands, of animals that apparently died in rivers.[136] However, such mass deaths could be the result of migrations enforced through unusual events such as a drought followed by a flood, rather than of normal seasonal migrations; though there is now strong evidence for the latter, too. Examination of isotopes in the enamel of dinosaur teeth reveal distinct and alternating signatures that point to animals alternating their time in different environments (see [137] on sauropods and [138] for hadrosaurs), indicating a regular movement between areas and, by inference, seasonal migrations. Others, though, were year-round residents, and some hadrosaurs have been shown to be perennial even in cold polar regions, based on their bone histology.[139] So, as ever, there will be high levels of variation between species, and probably even between different populations of single species, as seen in, e.g., extant wildebeest (*Connochaetes taurinus*), which may form mass aggregations as part of migrations in the north of their range but not in the south,[140] and some individuals may move little or stick to small home ranges.[141]

Collectively, these types of data provide a useful starting point for determining which regions dinosaurs habitually occupied, and by extension can feed into interpreting their behavior. For example, herbivores that lived in forested areas would presumably have found most available plant material off the floor, when those in open areas would have found most food lower down. Differing cover would have influenced communication (was line of sight clear or not?—see [142] on sensory ecology and signals) and ability to hide from predators, and would potentially have influenced behaviors from foraging and vigilance to migration patterns and nesting opportunities (see Case Study). The spatiotemporal variation of habitats can have profound effects on the ecology of animals and influence their motility, social biology, activity patterns, and more,[141] and so should not be ignored (if it can be determined).

Physiology

The physiology of dinosaurs is generally a complex problem, and a considerable volume of research has been dedicated to determining the likely metabolic rates of various species and clades. In particular, growth rates typically show rapid development and increase in mass that supports both a high and a stable body temperature (see [115]). There would inevitably have been considerable variation across the Mesozoic, but the relatively recent suggestion that mesothermy (intermediate physiology between the classic "hot" and "cold" split) was common in the Dinosauria has been strongly challenged (see [143] and responses).

Increasingly, it appears that the ancestral state of dinosaurs was that of endothermy (internally generated high temperatures), and also that feathers (providing insulation) were the ancestral condition in the Triassic. This is proposed for theropods, at least, but also for dinosaurs in general, with a secondary loss of filaments in both sauropodomorphs and some (or most) ornithischians. Modeling of the potential basal metabolic rate against various microclimates suggested that early small dinosaurs must have been insulated and maintained a high body temperature.[144] Similarly, Paul Olsen and colleagues[145] showed that early dinosaurs were primarily occupying regions in high latitudes that were both cold and potentially inhospitable to non-dinosaurian reptiles, suggesting again that their ability to generate and retain heat was the original condition for dinosaurs and that they were endothermic to some degree.

Such a high physiology does not automatically correlate with a large brain or higher intelligence, as has sometimes been suggested (given both the intelligence of some reptiles and the lack thereof in some birds), but it would have other important implications for behavior. As endotherms, dinosaurs would have had higher activity levels than most other Mesozoic terrestrial animals, and they had the potential to be active in all different weathers and climates (and as noted above, hadrosaurs are recorded as living though polar winters). It may also have been an exaptation (a preadaptation to later evolutionary events) and facilitated such high-energy activities as long-distance running and powered flight in various groups. Endothermy would, however, require higher food intake, increasing foraging times and determining important aspects of resource acquisition.

Posture

Dinosaurs were upright animals, with the legs held below the body and with a parasagittal (non-sprawling) gait. This posture and the fundamental carriage and organization of the limbs of dinosaurs provide a foundation for investigating their biomechanics (see below) and would have influenced their behaviors.

The neck shape of dinosaurs (and indeed of amniotes as a whole) is one that is fundamentally S-shaped, and so the head would have generally been held high.[146] Various early theropods had long necks, as did early sauropodomorphs, but this point is especially important for the very long-necked sauropods and their ability to reach up into the trees to forage[146] (see also Feeding).

The ranges of motion of joints across the skeleton have been investigated for some species, which allows us to interpret how far they could stretch or rotate various limbs or anatomical units, whether they could rear bipedally, and similar actions (see [147, 148]). In particular, there has been much work of this kind on the forelimbs of theropods, to assess their possible role in predation (see [149]), though this has yet to be extended to other activities.

In terms of being at rest, there is evidence of the postures adopted by some dinosaurs when not moving. One well-preserved trace of a Jurassic theropod shows the metatarsals of the foot in full contact with the ground in a plantigrade posture (very similar to that adopted by some modern birds at rest). Notably, there is a midline impression that is from the pubis resting on the ground and imprints of the ventral margin on both hands, showing that these were not pronated (i.e., the palms were not laid flat against the substrate).[150] Such a posture was likely common in theropods, at least, and partly matches that also seen in some body fossils of animals apparently in a resting posture when buried (figure 3.3). Several theropods and small ornithischians are preserved in what has been considered a "sleeping posture" (e.g., the small troodontid *Mei*[151]—see also figure 2.3): the animal crouched down with the head tucked to one side and in at least some cases, the tail wrapped around the body. While this is clearly a natural posture adopted by the animals, these are perhaps better considered to be "at rest" traces,[152] since the exact state of the animal is not known, and it could well be alert or at least awake rather than sleeping.

As fully upright animals, dinosaurs would be well clear of the ground when walking; their erect gait would be more efficient than that of sprawling reptiles, and they would be able to take longer strides. This advantage would extend well beyond their biomechanical performance (see below) to affect their general biology as well. Supported clear of the ground, even small taxa could, for example, avoid radiated heat from a surface that a small sprawling reptile would absorb, and this would likely have enabled dinosaurs to be active at times and in places that those others could not. Similarly, a generally upright posture would allow an animal to reach higher than a comparably sized sprawler, potentially giving access to new resources (especially for herbivores) and better vision of the surrounding area. Tucked resting postures would facilitate the retention of heat, and again might be important for animals that had a high basal metabolic rate and/or were endothermic.

FIGURE 3.3 Hypothesized resting posture for *Tyrannosaurus*. While the possible range of postures for giant theropods has yet to be explored, this is based on a known resting trace of a small theropod from the Jurassic, based on [150]. Artwork by Gabriel Ugueto.

Locomotion

In conjunction with endothermy, the upright posture of dinosaurs would make them generally comparable to birds and mammals in terms of biomechanics and performance. Quite how they moved, and what they were, and were not, capable of in various situations, is clearly a core part of a dinosaur's behavioral repertoire.

Most fundamental for terrestrial tetrapods is the ability to walk and run, and although there has been work on the ability of various dinosaurs to climb, dig, swim, and even fly, there has been much less research in these most basic areas. This is changing as ever more detailed studies of the bones, muscle insertion points, muscle mass, and joint flexion of dinosaur limbs allow for detailed studies of the locomotory capabilities of some animals (see [153]) (figure 3.4). For example, *Tyrannosaurus* has been shown to be relatively slow-moving compared to many early estimates,[154] though it was an efficient mover and one that was specialized for moving long distances effectively,[116] with obvious implications for its behavior as a carnivore.

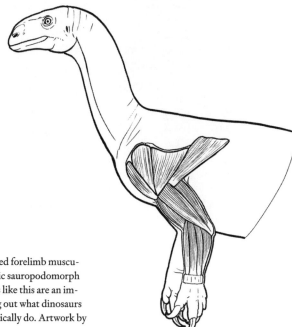

FIGURE 3.4 Reconstructed forelimb musculature for the Late Triassic sauropodomorph *Thecodontosaurus*. Studies like this are an important basis for working out what dinosaurs could and could not physically do. Artwork by Gabriel Ugueto.

FIGURE 3.5 Hypothesized posture for the small gliding dinosaur *Microraptor*, shown climbing a tree. Exactly what the range of motion was for the arms and legs, and how these could be used to move up and down or along trees, is not known. Artwork by Gabriel Ugueto.

Small dinosaurs probably could climb, and in particular small paravians (the theropods closest to birds) were likely well suited to this behavior.[28] Although even these dinosaurs do not show obvious adaptations for arboreal activities, like hypermobile joints in the wrists and ankles or enlarged reversed toes for gripping branches, they nonetheless exhibit a suite of characters that would assist them in getting into and moving around in trees, such as curved claws with a strong grip which would have facilitated climbing[28] (figure 3.5).

At least some small and feathered theropods must have been active in the trees, because of the strong evidence of their ability to glide (e.g., the small dromaeosaur *Microraptor*[7]). Although their exact capacity to fly is uncertain, the presence of enlarged and asymmetric feathers on the wings points to animals that could move through the air. Even if they were restricted to passive gliding, climbing must have been a key part of their lifestyle to gain height between launches. Recently it has been suggested that several of the most birdlike dinosaurs may even have been capable of powered flight,[30] which would have opened up other possibilities for behaviors in these animals (and in particular, limited the use of trees as a launch point[155]), though this is both uncertain and currently confined to only a handful of species.

Many dinosaurs would also have manipulated their environment and been able to dig. This would likely have been a possibility for many species, even those not obviously adapted for digging (as seen, for example, in theropod courtship scrapes[35]), though a number of clades across all three major groups do have features that might have facilitated this. Alvarezsaurian theropods demonstrate adaptations for digging into substrate,[156] though they may not have dug holes or burrows; the enlarged thumb claw of many sauropods would have been effective in digging and might have been important for making nests.[157] More notably, the small thescelosaur *Oryctodromeus* both has been found preserved inside burrows and also shows adaptations in the forelimbs consistent with digging, suggesting that it was responsible for these excavations.[29, 158]

Almost all tetrapods can swim and propel themselves in water (even if many are very poor at this), and there is similarly good evidence for dinosaurs moving in water, most notably from various trackways (see [82]). Studies have shown that most dinosaurs should have been capable of at least maintaining stability when floating in water, which would allow them to propel themselves; and rigorously produced models yield floating postures that match some trackways well (see [26]). Very few dinosaurs, however, are thought to have been adept in the water; though the recently discovered halszkaraptorines (an early branch of the dromaeosaurs) appear to show adaptations toward swimming,[159] this has also been challenged. (See also Case Study on *Spinosaurus*.)

The abilities of animals to move around in (and, as in the case of digging, manipulate) their environments have profound implications for their behaviors. The acceleration and speed of both predators and prey are one obvious dynamic, and abilities (or otherwise) of animals to climb, fly, or swim will obviously limit or expand their niches and behavioral repertoires accordingly. Especially in larger taxa, locomotory mechanics may change profoundly between juveniles and adults, such that large size may, for example, limit top speed while also enabling adults to wade into deeper waters or cross other obstacles.

Summary

Collectively, this data gives us a strong base for analyzing dinosaur behavior, comprising the basics of what is likely possible and plausible for many essential aspects of their activities. The following chapters move into the details of key areas of dinosaur behavior, and while many of the examples discussed there will be focused on single lines of evidence, these should be considered against a backdrop of the kinds of studies and types of data outlined here. Analyses of locomotion or activity patterns, for example, may not feature in discussions of plant choice by a large sauropod, but when building a picture of the complete behavior of a species, it would be a consideration and should not be overlooked. Often these areas are not well connected, but as studies increase and evidence builds, it should become increasingly common to be able to interpret and contextualize the behaviors of species in a holistic way that integrates ideas and tested hypotheses across numerous sources.

CASE STUDY: The Aquatic Affinities of *Spinosaurus*

Establishing even the basic habits of dinosaurs can be difficult and controversial. Limited fossils and apparently contradictory results of analyses, or conflicting interpretations of the available data, are not uncommon. Even when it is clear that, say, animals must have been spending considerable amounts of time in and around water because fish were a major part of their diet—how much time they spent there, whether or not they could swim well, and what else they might have eaten are all pertinent questions and can be investigated.

The giant North African megalosauroid theropod *Spinosaurus* has long been problematic to discuss thanks to the destruction of the original (limited) remains in World War II. It was obviously a large animal with rather crocodile-like teeth in a long jaw (figure 3.6) and a huge sail of elongate neural spines

FIGURE 3.6 A very large snout (premaxilla and maxilla) of *Spinosaurus*, showing the long jaws and some of the conical teeth. Specimen MSNM V4047 at the Museo di Storia Naturale di Milano. Photograph courtesy of Cristiano Dal Sasso.

along the dorsal series of vertebrae, but not much else was clear. Subsequent discoveries of other spinosaurs in the 1980s and 1990s greatly added to our understanding of the group and posited them as large-bodied animals with crocodile-like jaws and teeth, large-clawed arms, and an at least partly piscivorous diet, based on the evidence of data from isotopes, gut content, and comparative anatomy of the skull.[160] But how aquatic were these animals really? Where were they spending their time and what were they doing?

New finds of an incomplete subadult *Spinosaurus* from Morocco shows that it had remarkably short legs, a fin-like tail, and unusually dense bones.[161-163] It has been described as "semi-aquatic,"[161] and later as an aquatic "pursuit predator"[162] and as a "subaqueous forager."[163] In short, it is considered to have spent considerable amounts of time in water and potentially to have even been highly specialized for the role of swimmer and diver.

However, critical appraisal of these and other lines of evidence used to support these hypotheses has found them wanting. Although dense bones would help the animal sink, the bones may not be as dense as originally suggested;[166] the skeleton remains pneumatic, and it would float, making diving energetically expensive.[164, 165] The fin-like tail would provide insufficient power to propel the animal at speed or downward,[165] and in morphology it more closely resembles the tails of animals that use them as signaling structures rather than to swim.[164] The size of the sail would also induce huge drag costs, further reducing efforts to swim,[164] and the position of the weight of the sail would make the animal unstable in water.[27, 165] The legs, while short, would allow the animal to move effectively on land,[27, 165] and some spinosaurs (though not all) may have had a habitual head-down posture (see [167]), well suited for striking at prey from above.

Perhaps most importantly, various North African spinosaur teeth show a mixture of isotopic signatures. Although some are closely aligned with those of crocodiles, turtles, and other animals fundamentally tied to the water, some others showed a signature that resembled those of theropods feeding purely on terrestrial sources of food.[168] Given that teeth are shed every few months, this would imply that these spinosaurs had spent sufficient time on land to have developed a fully terrestrial signature for one round of teeth, at least. Even if they were mostly spending their time in water, some spent considerable periods of time out of the water.

In short, despite the apparent aquatic affinities of the animal, *Spinosaurus* was probably not highly aquatic or a specialist swimmer. It did not have a greatly reduced terrestrial capacity, and spent time on land, both to get food and also likely to move between bodies of water, where it hunted for fish and other prey. It is perhaps best interpreted as something akin to a giant wading bird (if a flightless one), taking assorted prey from both land and water[164] (though even this is probably not the whole story)—more aquatic than other theropods, but not a form of theropodan crocodile. This conclusion is reached only by assessing numerous lines of data and still is open to other possibilities if, for example, there turn out to be multiple distinct populations or even species across North Africa that could explain some of the apparently conflicting signals. The perhaps inevitable conclusion is that more data is needed.

CHAPTER 4

Group Living

In this chapter we engage with the complicated issue of dinosaurs living in groups. Various dinosaur species and clades have been argued to have lived in groups, with some even suggested to have actively cooperated in what would presumably have been complex structured social groups that hunted together (see Case Study). The idea that dinosaurs may have had complicated social structures and shown various degrees of gregarious behavior and group living is not a new one,[169, 170] though it remains one that has perhaps been overstated and requires careful consideration (see [33]).

One problem with the discussion of social behavior in dinosaurs is that multiple different terms have been used interchangeably without a clear definition or diagnosis, leading to confusion, or at least a lack of clarity (though this is hardly an issue restricted to dinosaur research; see [98]). Here I use the term "aggregation" to relate to any grouping of animals for any reason (even a pair of an antisocial species that are together only to mate); "gregarious" is used to denote that members of a species have a tendency to congregate for various reasons (mutual protection, migrations, foraging, mating opportunities, etc.) and tolerate one another, without necessarily maintaining any formal groupings (such as family groups or harems); and "social" animals are those that fundamentally live in groups as a major part of their lives, and do not typically live alone; these would also generally have some kind of hierarchy and social structures in the group (figure 4.1).

Living in groups can be driven by a number of different factors and can operate at multiple different levels, depending on the biology of the animals and their environments (see [171] for an introduction to the subject). The formation of groups can especially be related to

FIGURE 4.1 A large herd of Cape buffalo (*Syncerus caffer*) in South Africa. Although there are more than 100 animals here, this is actually a group of multiple smaller herds that travel as a larger unit. Photograph by the author.

age, since, for example, adults need to find a mate and juveniles are more vulnerable to predators; each may require different resources in different amounts, driving different behaviors and habitat occupation (patchy resources will also inevitably force members of a species to the same areas[172]). Individuals may therefore readily switch between differing levels of sociality based on their age, sex, social status, reproductive state, and more (see [92]).

One major reason for animals to come together is defense. Herbivores in particular may need to spend a very long time finding food and feeding, and this reduces their vigilance. Living in a group promotes defense of each individual, increasing the chances that one member of the group will detect a predator (the vigilance effect). Group size (or even the aggregation of otherwise solitary animals) may be driven by the density of predators (see [173,174]), and a group may be able to fend off threats that single animals cannot. If attacked, groups can also offer protection though the dilution effect: if numerous animals are nearby, any individual is much less likely to be targeted.[175] Groups can additionally increase foraging efficiency, though they can also generate increased competition for the resources when these are found.[176]

FIGURE 4.2 One driver of aggregations is foraging success. Here a family of southern ground horn-bills (*Bucorvus leadbeateri*) are hunting in a savannah; any animal flushed out by one may be caught by another group member. These animals are also cooperative breeders with older juveniles helping their parents. Photograph by the author.

Coming together in a group can also provide sufficient numbers to defend a territory or resource. For example, in lions (*Panthera leo*), Anna Mosser and Craig Packer[177] demonstrated that larger prides controlled better territories and their members had higher fitness, rearing more offspring than those of smaller prides. Control of re-sources is a critical factor that has been suggested to drive cooperative breeding in particular. The variations of environments in both space and time can also create conditions where cooperation (especially be-tween generations) in breeding will increase the fitness of individuals and promote cooperation (see [178]).

With greater social interactions may come cooperation, thus allowing for mutual benefits in terms of effort in rearing offspring, though this may be a complex subject with both benefits and costs (e.g., as seen in acorn woodpeckers, *Melanerpes formicivorus*[179]). The most obvious trade-off here is that aggregating in a group leads to increased competition with conspecifics with which an individual will have the most similar resource requirements. As a result, larger groups may have to spend longer feeding and move farther than smaller groups.[180] Aggregations may also occur for entirely sepatare reasons, such as a single resource or draw for normally solitary species, such as snakes aggregating to hibernate.[181]

Social groups may be composed of closely related animals, especially when these are juvenile-only groups or adults with a cluster of offspring, but not necessarily. Still, it is impossible to know the relationships of individuals in a fossil aggregation, so terms such as "family group" should generally be avoided. (Nests may of course be an exception, but even here, the possibility of communal nesting and care of young could mean that nestlings are only half siblings, or even unrelated, and an associated adult may not be a genetic parent to any of them.) At the other end of the spectrum, aggregations may not even be limited to a single species. Mutual benefits can lead to collaborations and even social interactions between different species (e.g., hornbills, *Bucorvus*, grooming warthogs, *Phacochoerus*[182]), even those that might normally be considered competitors. For example, two or more species of herbivores may aggregate, offering mutual vigilance and defense against predators as well as potential foraging advantages (see [183]), and perhaps these associations form especially

FIGURE 4.3 A mixed aggregation of ungulates (wildebeest, *Connochaetes taurinus*; impala, *Aepyceros melampus*; plains zebra, *Equus quagga*). These groups may form temporarily because of a combination of factors, such as a limited resource (e.g., water), access to a plentiful resource with low competition (e.g., grasses), or protection from predation (increased vigilance and the dilution effect). Photograph by the author.

where they have complementary senses and do not compete for the same resources (figure 4.3). But interspecies cooperation in hunting has been seen between wolves (*Canis lupus*) and striped hyena (*Hyaena hyaena*),[184] so cross-species cooperation is not limited to herbivorous taxa.

Group living can also have other effects too, such as the positive correlation of group size and cognitive performance seen in some birds;[185] notably, brain size can correlate with social behaviors in ungulates[186] and birds.[187] Some hadrosaurs have been noted to have larger brains; that might point to their having lifestyles more social than those of their relatives[69] (see also chapter 5, Signaling). These data suggest both obvious potential advantages to living in a group, and also a potential line of evidence for inferring degrees of sociality in dinosaurs.

At least some dinosaurs must have lived in groups throughout their lives; for others, sociality would have varied as they grew or as circumstances dictated (e.g., droughts can force groups toward water sources or migration to less depleted areas), but despite the

abundance of dinosaur fossils found in groups, the actual evidence for group living is rather sparse. Fossils capture only a moment in time, and whether or not this represents what was normal may be difficult to determine. Degrees of social interaction, and behaviors such as dominance or cooperative rearing, may be impossible to determine.

Problems with the Evidence

Dinosaurs would certainly have had the capacity to form aggregations, and even social groups. Birds are often highly gregarious and even social, and while crocodilians are not normally gregarious, they may aggregate in large numbers with little antagonism, especially in their breeding seasons.[188] As large carnivores, however, crocodilians may not be an appropriate model for dinosaur social behavior, although notably, among other reptiles some iguanas (e.g., *Iguana iguana*) display extensive, complex social behaviors. Here juveniles may engage in social interactions including allogrooming, and can even show leadership-like behaviors.[181] Similarly, there is evidence for various social behaviors, from hunting to predator avoidance, in some other extant reptiles as well (see [105]).

Direct evidence for dinosaurs living in groups comes from abundant body fossils and mass mortality sites of numerous animals,[189] and from trackways which show multiple individuals of the same species moving together in the same direction.[190] Both these lines of evidence naturally come with caveats, which means that fossils cannot immediately be taken at face value, and the context in which these sites were made, and their taphonomic history, is important—dinosaur fossils found in groups did not necessarily live, or even die, together.[191] In the case of body fossils, animals might have been driven together out of necessity, such as by a shortage of resources, or been killed when they were together for only a brief part of their lives. Grizzly bears (*Ursus arctos horribilis*) are typically antisocial, but will gather in large numbers to feed on a glut of fish or a whale carcass; they could potentially be killed and buried together by a storm surge, resulting in an apparent "social" group being preserved together. Similarly, animals might aggregate for seasonal migrations to better conditions, but risk being killed by crossing rivers or similar hazards. This would happen annually and could result in a huge buildup of bodies, potentially at multiple sites, but again, it would be misleading to suggest that these animals normally lived in groups.

Data from tracks present similar issues, as animals might be drawn to a single location for reasons such as a lek (males gathering to compete for females), or a limited water supply, and leave tracks suggesting large numbers of animals were together (at least one case has been inferred for dinosaurs[192]). Some tracks have been shown to follow paleo-shorelines, for example, and so demonstrate parallel tracks of apparent gregarious behavior that might simply result from the limitations of local geography (see [193]). Evidence as slight as tracks of two individuals at one site has been used to suggest that a whole clade might be social, when this could be anything from a bonded pair to a chance encounter, a brief coming-together in the mating season, or even a rival trying to drive another off. Moreover, without evidence that both sets of tracks were laid together (for example, animal A leaving prints over those of animal B and vice versa), it is also possibile that the tracks were laid down hours or even days apart and that the two individuals never even met.

Identifying taxa from trackways is also often difficult (see [82]), and so aligning tracks and bonebeds, or even multiple trackways, and assigning them to the same genus or species is also very difficult. Annette Richter and Annina Böhme[194] provide a nice example of a tracksite showing large numbers of densely packed tracks of multiple different species, including ornithopods, ceratopsians, and both large and small theropods (the latter, at least, were predatory taxa), traveling in multiple different directions across a deltaic area. Differential directional trends of different track types show that different animals were doing different things for probably different reasons, so determining what prompted these tracks to be laid down in these patterns is clearly a challenge—and all the more so for more isolated ones (figure 4.4).

As before, multiple lines of evidence and multiple independent samples are most important for inferring that animals were spending

FIGURE 4.4 A mixed group of species drawn together by a single resource, but not engaging in social interactions. Above, several species of large birds scavenge on a dead antelope (left to right: hooded vulture, *Necrosyrtes monachus*; white-backed vulture, *Gyps africanus*; marabou stork, *Leptoptilos crumenifer*). Similarly (below), these tracks are left on a game trail, and although multiple animals of different species are using this path, there is no reason to think they were interacting. Photographs by the author.

time together. Some dinosaurs were truly gregarious, but which, and to what degree, and when in their lives, and under what conditions is mostly uncertain. The overstatement of the inference that a cluster of individuals were social is obviously therefore problematic. On the other hand, circumstances for preserving animals in groups must also be interpreted correctly—doubtless large numbers of species lived their entire lives in groups, but mostly those would die one at a time, losing single animals. If the herds avoided muddy areas that might leave traces, then they would leave no evidence that they were gregarious, but would leave multiple individual bodies—which might be interpreted as solitary. Thus, while we should be cautious about overinterpreting mass mortality sites as evidence of dinosaur sociality, similarly the absence of such clusters does not rule out social behaviors.

Evidence of Dinosaurs in Groups

Body fossils or tracks potentially suggesting sociality have been recovered that are attributable to numerous different dinosaur clades across the sauropodomorphs,[32] theropods,[195] and ornithischians.[196] Some of these finds involve numbers of animals that could be in their thousands (the ceratopsian *Centrosaurus*[91]) or down to a single pair, apparently a male and female (the oviraptorosaur *Khaan*[197]). These finds are also extremely varied in composition, and may consist entirely of adults (see [198]), of all juveniles ([199, 200]), or mixed groups with multiple different life stages represented ([120, 201]).

In the case of theropods, numerous sites preserve either monospecific bonebeds or trackways showing multiple individuals moving together. These cover a wide variety of taxa, both in phylogenetic position and in gross morphology and behavior, from small herbivores up to giant hypercarnivores. The largest such find, and one of the most important, is of the small Triassic carnivore *Coelophysis*, known from an aggregation in New Mexico that includes the remains of several hundred individuals[195] (figure 4.5). Although several other small theropod taxa have been named from the assemblage, it is dominated by similar-sized (and apparently similar-aged) animals and suggests an enormous aggregation of *Coelophysis*. Other aggregations of adult skeletons of (probably herbivorous) adult oviraptorosaurs are also known.[202] There are numerous trackways of small theropods (which may represent juveniles; see [203]), but some are associated with tracks referred to the carnivorous abelisaurs,[190] which are of large animals and therefore presumably adults. Importantly, there are also skeletal aggregations that have been used to infer social group hunting for

a

10 cm

FIGURE 4.5 Several individuals of the small Late Triassic theropod *Coelophysis*. These are part of a huge aggregation of over 100 individuals known from a single site in New Mexico. Photograph courtesy of Danny Barta.

both dromaeosaurs[204] (and see Case Study) and large tyrannosaurs. In the case of the latter, it has been proposed that these habitually lived in mixed-age groups.[120] While these clusters feature multiple examples of some species (notably *Albertosaurus*), and accumulations over time have been ruled out,[120] the extent of their time together and degree of cooperation remain uncertain.[33]

A number of early sauropodomorphs are known from monospecific bonebeds representing mass mortality sites. *Plateosaurus* from the Late Triassic is known in huge numbers of specimens across multiple sites in Germany, each of which contains multiple animals,[205] and these are typically all large individuals. The Early Jurassic *Mussaurus* from southern Argentina, in contrast, is represented by specimens from eggs through to adults, but at least some sites show age segregation,[201] demonstrating that this is not just a nesting ground, but represents multiple groups of animals.

Sauropod body fossils are less common as monospecific bonebeds, although some are known (see [206]), and others are dominated by sauropods even though other taxa are present (see [207] on the Late Jurassic Morrison Formation). Evidence for aggregations of sauropods is more typically found in the ichnological (trace fossil) record, and multiple good trackways are known, covering a wide variety of clades.[32, 208] Across both bonebeds and trackways, there is evidence both for age-segregated groups and for those with adults and young animals preserved together.[200]

One interesting aspect of sauropod groups is the sheer size of the animals. One disadvantage of aggregations of a species is that individuals have similar resource requirements, and most sauropods would need to consume a lot of food, which might lead to high intraspecific competition. Given the frequent suggestion that large sauropods would be largely immune to predation, groups might not add protection benefits, either (though see [6] for a contrary position, and see also chapter 8, Feeding), making aggregations in this clade a possible outlier.

The best data for aggregations comes from the ornithischians. This may reflect that they are more common than theropods, and much smaller than sauropods, so that compared to the latter, they would potentially be more vulnerable to predators and would more readily form groups—and could more easily become trapped or buried to become preserved. Numerous records of ornithischian monospecific bonebeds are known, representing almost every major clade. Among early-branching ornithischian groups this includes the small *Kulindadromeus* from Russia,[19] and among armored dinosaurs, groups of ankylosaurs are known that include both adult-only and juvenile-only groups.[198] However, among the more derived ornithischians of the Cretaceous, such records are especially common, with numerous monospecific bonebeds for the ornithopods ([109, 196, 209]) and ceratopsians ([21, 31, 189]). Similarly, for these same groups there are multiple mass tracksites with tracks of numerous similar-sized individuals preserved together,[210] which also preserve evidence of herd structures within these groups (in the case of hadrosaurs[211]).

The ceratopsids in particular have been a focus of study. Some enormous groups are known that contain thousands of fossils and at least hundreds of individuals, and many therefore appeared to have lived in groups for at least part of the year.[172] Based on the taphon-omy of the sites, at least some of these aggregations do not appear to have been stress-induced (e.g., by drought), while many others are equivocal at best.[212]

Juveniles

Notably, there are examples of aggregations of juvenile dinosaurs, even when adults of the same species are normally found alone (e.g., the large neoceratopsian *Triceratops*[213]), while other sites show aggregations separated by age into multiple size classes, from young juveniles through to adults (e.g., the small ceratopsian *Protoceratops*[214]). This pattern is again widespread, and includes ornithomimiosaurs,[215] early sauropodomorphs,[201] diplodocines,[216] ornithopods,[217] ceratop-sians,[218] and ankylosaurians.[198] There is often some uncertainty about the exact age or life stage of non-adult dinosaurs, although here the term is restricted to those that are likely below breeding age (i.e., it does not include animals often referred to as "subadults") and also not the very youngest animals (hatchlings or neonates) that would be associated as a result of being in a nest.

Younger dinosaurs, which would be smaller and weaker than adults, might have been more vulnerable to being trapped and bur-ied, e.g., in mudflats, which could influence this pattern.[214] However, the presence of groups of juveniles with adults (see above) means that this cannot always be the case, and many incidences of preserved juvenile-only clusters must be from incidences where adults were not present. Indeed, the records of all-juvenile dinosaur bonebeds are common enough (again, some include multiple examples of single species) that many young dinosaurs must have spent considerable time together to be captured in this way[199] (see also chapter 6, Re-production) (figure 4.6).

The drivers for juvenile gregariousness are clearer than those for adults. If groups were migrating, were driven together by an issue such as drought, or were being cared for by parents, we would ex-pect to find mixed-age aggregations; and juveniles would not come

FIGURE 4.6 A group of juvenile *Sinornithosaurus* trapped in mud and unable to escape. This is based on the aggregation of these animals described in [199]. Artwork by Gabriel Ugueto.

together to breed. Therefore, their drive to aggregate, but under a different pressure than adults, points most clearly to their vulnerability to predation. Juveniles are preferred as prey by predators, and there is evidence for thisin dinosaurs (see also chapter 8, Feeding). Juveniles are naïve about predators, and spending more time foraging also makes them vulnerable to predation; adults may even keep juveniles away from them to reduce competition and to avoid attracting predators (see [88]). Furthermore, juveniles often lack the defenses of adults, whether that is horns or other offensive weapons seen in various groups, or even armor: juvenile ankylosaurs notably have small osteoderms, or none at all.[219]

Summary

Ultimately, there are very large numbers of tracksites and bonebeds that show various dinosaurs moving together in large aggregations or being together at the time that they died. In some cases, there are multiple examples of the same species being recovered together in groups, showing that these animals must have been spending considerable time together in aggregations and were likely at least gregarious, and potentially social. This would still not rule out the possibility that other individuals or populations were more solitary, or that this general gregariousness was seasonal or situationally dependent, but clearly, groupings cannot have always been the exception.

Just how social these animals were, and whether they engaged in social interactions (e.g., social dominance and behaviors like collective hunting or allogrooming), is impossible to determine from the fossil record, though other data may yet arise. Certainly the presence of numerous age-segregated groups points to differing ecological pressures on different age groups, and perhaps integration of data on body size with prey preferences of theropods, color patterns (for camouflage), diet, and habitat occupation could help generate holistic patterns of likely drivers of gregarious vs. solitary living for some species. Coupled with a better understanding of the general ecology of dinosaurs, and of the specific circumstances that can drive group living in extant animals, it may well be possible to better assay the presence and evolution of these behaviors in the fossil record. However, the limitations of bonebeds (relatively rare) and trackways (hard to assign to specific genera) make this a difficult task. Juveniles, at least, were likely gregarious based on their strong vulnerability to predation, which would have made shared vigilance and the dilution effect important.

CASE STUDY: Evidence for Pack Hunting in Dromaeosaurs

One of the more contentious examples of group living in dinosaurs is that of cooperative hunting in theropods, most notably the dromaeosaurs. This is an idea that has sunk into the public consciousness,[220] and certainly "pack hunting" by these animals is a common theme in books, documentaries, and artwork. The actual support for this behavior is extremely limited, however, bedeviled with alternate interpretations of limited data, and, inevitably, based on a single taxon and then blanket-applied to the clade as a whole.

The center of this work is the large (for a dromaeosaur) *Deinonychus* from the Early Cretaceous Aptian-Albian Cloverly Formation of the western United States. This discovery of this genus was an important part of the recognition that birds evolved from dinosaurs, but it was also interesting as this included the discovery of multiple individuals preserved together in association with a skeleton of the early ornithopod *Tenontosaurus*. It was suggested that this was evidence that a group of these dromaeosaurs had hunted together to bring down the large prey animal, in the manner of modern canids.[170] *Deinonychus* remains (especially shed teeth) are known from around a third of the sites that also contained *Tenontosaurus,* suggesting the possibility of a genuine ecological relationship between the two,[221] although this association is perhaps not surprising as the herbivore is the most common dinosaur in the formation. Of these fossil localities, he most important for hypotheses of coordinated predation is one site with the remains of at least four partial *Deinonychus* skeletons in association with that of a single *Tenontosaurus* (figure 4.7).

Subsequently, Brian Roach and Daniel Brinkman[220] provided a detailed response to the arguments that *Deinonychus* operated in coordinated groups to target large prey. A major part of their counterargument was the lack of evidence for coordinated hunting of large prey by extant archosaurs as per the extant phylogenetic bracket. There are, though, at least some hunting birds that do cooperate (see [34, 220]) when hunting small prey; and given that the highly derived nature of birds, their relatively small size, and their ability to fly means they have very different ecologies than terrestrial theropods, this might not be a fair comparison, with the ecological limits of flight placed on birds that would not be present on terrestrial carnivores. So perhaps it is not unreasonable to infer that group hunting could have evolved in dinosaurs, despite the EPB.

More importantly, large prey certainly can be killed by single predators, as Roach and Brinkman note for Komodo dragons (*Varanus komodoensis*), and lynx (*Lynx lynx*) are documented to actively hunt and kill adult red deer (*Cervus elaphus*),[222]

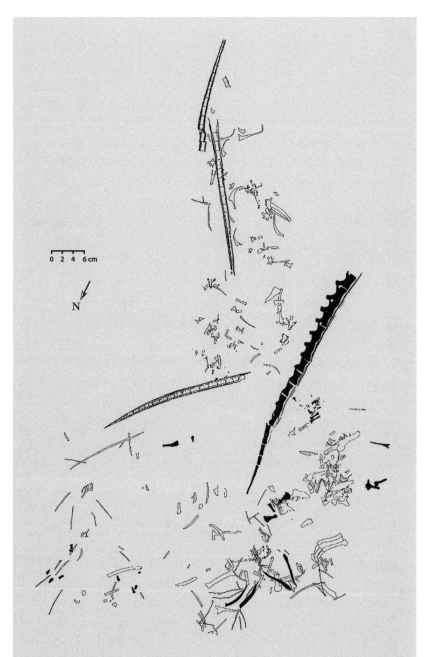

FIGURE 4.7 Quarry map of the *Deinonychus–Tenontosaurus* association that was used to infer group hunting in dromaeosaurs. The bones of *Deinonychus* are in white while the *Tenontosaurus* bones are colored black. Figure modified from the original by Maxwell and Ostrom.[221]

demonstrating that lone predators can and do tackle prey multiple times heavier than themselves. Furthermore, while adult *Tenontosaurus* were large animals, most of those noted as having possibly been killed by *Deinonychus* were only about half-grown, and so could at least potentially have been tackled by a single dromaeosaur.

Roach and Brinkman[220] suggested that in fact these animals were not cooperating, and that multiple casualties alongside a dead prey animal represented evidence of *Deinonychus* fighting and killing each other, rather than a *Tenontosaurus* somehow killing most of a pack of hunters. They point again to Komodo dragons, which can bring down large prey, and then subsequent fights over access to the carcass can lead to deaths and cannibalism. The presence of one *Deinonychus* with the claw of another stuck in its tail adds weight to the idea of intraspecific aggression.[220]

As conceded by Desmond Maxwell and John Ostrom,[221] there is no definitive evidence to show that the *Tenontosaurus* was killed rather than scavenged. Certainly it seems unlikely that a mid-size iguanodontid would be able to kill multiple predators that could enact a coordinated attack. If that was normal, then it does not speak well for the abilities of *Deinonychus* as an effective predator. On the other hand, if that was not normal, then this assemblage should not be taken as strong evidence for pack hunting of large prey. Thus, either way, this association does not make a strong case for social and coordinated hunting.

Only one other dromaeosaur is known from sites representing multiple individuals (and this may be a predator trap[204]), so the idea that these animals as a clade are pack hunters is essentially based on one genus alone, and the evidence for that is at least questionable and likely incorrect. Roach and Brinkman noted that the overinterpretation of theropods in general as pack hunters was based on sparse evidence and possibly influenced by John Ostrom's[170] assertions about *Deinonychus*. Although they considered pack hunting unlikely for any theropod (an argument that also seems incorrect across the vast span of dinosaur history, and evidence for which would be very difficult to find), this case shows the danger of overinterpreting limited data and the importance of bringing together multiple lines of evidence from extant animals and careful assessments of fossil data (including taphonomy). We are left with a frustrating lack of clarity.

CHAPTER 5

Signaling

Here the focus is on various forms of communication and signaling within and between dinosaur species. This includes "positive" signals, such as dominance and sexual identifiers; "negative" signals such as aposematic (threat or warning) displays; and also "anti-signals," in the sense that camouflage is an attempt to *not* signal an animal's presence. Dinosaurs would undoubtedly have communicated with one another (especially conspecifics) in different ways; there is evidence of visual, auditory, and tactile signals or communication in various clades, and olfaction would surely have been used in others (figure 5.1). The recent discovery of fossilized melanosomes, allowing the identification of aspects of dinosaur patterns and colors (see [9]), has opened up new areas of work in dinosaur communication, though this line of research is currently limited by the rarity of appropriately preserved specimens.

The evolutionary drivers of signals vary with the information that the two organisms concerned intend to give and receive (see [223] for an introduction to signals). This information may be in the form of

FIGURE 5.1 The early tyrannosauroid *Guanlong* from the Middle Jurassic of China. Here reconstructed with two versions of soft-tissue integuments, one with the minimum associated with the bony head crest, the second with numerous elaborate expansions such as an inflatable dewlap, elongate spar-like feathers, tubercles on the face, and more. There are no osteological correlates for any of these and so no evidence they were present, but they are known in various other dinosaurs or birds and remain at least possible, though unknown. Artwork by Gabriel Ugueto.

"signals" or "signs" that have evolved for communicating information to another animal, or of "cues," such as tracks, which have not evolved for that purpose but may still communicate information[224]—though not all researchers follow this division (see [223]). Here, though, the interest is specifically in evolved signaling traits, although aspects of cues are also relevant in places.

Signals may be honest, such as the size of a horn that accurately reflects the ability of the bearer to grow it, or dishonest, such as an animal puffing itself up to appear larger than it really is. These can both become part of a Red Queen arms race, producing ever larger and more elaborate signals, and/or ever more complex forms of testing and detecting dishonest signals. Honesty of signals is linked to the "intent" (i.e., selected benefits) of the sender and receiver (see [225]). For example, two members of a social group would both benefit from being able to be able to identify each other correctly, so there is no driver for dishonesty, and moreover, no need for a costly signal (such as a large and elaborate crest) in order to advertise or confirm identity. In such a case, a simple and low- (or zero-) cost signal such as a unique call, scent, or color pattern would likely evolve. However, in a contest over resources, there would be a strong selection pressure for any animal that could falsely display dominance or seniority, leading to the evolution of honest signals such as a crest or bright pattern that was costly to produce or maintain (cost could be the weight of the feature, or that fact that it makes the animal more prominent to predators, etc.). Honesty also functions both intra- and interspecifically: for example, the "stotting" or "pronking" jumps of antelopes and other ungulates (in which all four feet are used simultaneously to jump off the ground while the head is held down in a stereotyped posture) are considered to be an honest signal of health to deter predators from pursuing them.[226] Signals may also be multifunctional, and an honest signal such as horn size, which would be used to advertise the quality of an animal to a potential mate, could also serve as a threat display to potential rivals or predators.

Determining which of these factors may have influenced the selection of various dinosaur signals is, inevitably, very difficult, and many of the selective pressures will be unknown, but some can at least be hypothesized and potentially tested. Common drivers are likely to be social or sexual dominance contests or advertisements, staying in contact with a group (be it a gregarious group or a parent and off-spring), and warning signals. Different signal types can communicate different information, but also operate differently in different environments and conditions. Most notably, visual signals (figure 5.2)

FIGURE 5.2 Elaborate signals are common in birds, both in large carnivores like the king vulture (*Sarcoramphus papa*) and in smaller herbivorous birds like Bulwer's pheasant (*Lophura bulweri*). Notably, many of these features would be very unlikely to preserve as fossils. Photographs by the author.

only operate in direct line of sight, and the receiver must be looking at the signaler to detect the signal (which could be reduced by low-light conditions). This may be an advantage, though, if the signal is concealed from all but the receiver. In contrast, auditory signals can operate across very long distances, can operate out of line of sight, and could be picked up by a passive audience. They may, however, function very differently in cluttered vs. open environments, where differing pitches and volumes may carry in very different ways. In addition to operating out of line of sight, olfactory signals are unique in that they may linger over a considerable period of time and can operate even in the absence of the signaler. There are, therefore, contrasting reasons why a given signal type may be favored over others, and, as ever, signals may have more than one function. Furthermore, a given signal may communicate different and even conflicting information to different individuals and be reliant on their attention and sensory ability to receive the signal (e.g., a scent may allow a male carnivore to advertise for a female while also potentially scaring off some prey, but not being perceived by others) (see [227] for a comprehensive review).

Sexual Selection and Dimorphism

Perhaps the most important aspect of signaling to consider is that of sexual selection, which can have profound effects on the evolution of lineages, and especially on their appearance.[228] Here animals are, in some way, advertising their quality as mates and their overall fitness so as to increase their reproductive output by securing more mates, better-quality mates, or both. Unlike natural selection, sexual selection always operates intraspecifically, although for some features is may be in harmony or in contrast with natural selection (e.g., it may favor horns that can also be used to fight off predators, or bright colors that make an individual more conspicuous to them). Sexually selected structures can take a variety of forms, and distinctions here are are important for the sake of clarity. Notably, Erin McCullough and colleagues[229] have argued that "weapons" that have evolved through (typically) male–male competition should not be confused with "ornaments" that arise through (typically) female choice. Both arise from reproductive competition, and both would be condition dependent (i.e., honest), and while weapons can certainly signal the quality of the animal, they ultimately serve to harm an opponent.[229] Here we are considering ornaments, while weapons are covered in chapter 7, Combat. Normally, sexual selection in terms of ornaments and weapons is centered on males, which bear the features that females do not, and/or are fundamentally larger animals. However, it should be stressed that this is not exclusive to males, and that females may be the larger and/or adorned sex. Notably, these patterns can be more complicated in females, for whom the development of other structures can link to the acquisition of resources in addition to mates.[230]

Definitions of sexual selection and its scope inevitably vary (see [231]), but the concept has been especially controversial when applied to dinosaurs, provoking, in particular, the assertion that it did not operate in these animals because of a lack of sexual dimorphism. Although dimorphism may be difficult to detect, and its absence does not rule out sexual selection (see below), it has been suggested that the various visual signals of dinosaurs were instead used for species recognition.[232] As a concept, species recognition may cover several different aspects of social behavior, from correct identification of potential mates to coherency in herds. In both cases there would be selection to reduce signal costs, and indeed to select for low- or zero-cost signals (e.g., a scent or a ritual action), as both sender and receiver require an honest signal.[225] High-cost signals in the form of large bony ornaments do not fit this pattern, and the pattern of development where these crests are absent in (often gregarious) juveniles but present in adults, some of which were solitary, does not support the argument for coherency.[233] Formal testing of the idea of species recognition by assessing if sympatric species were more different than allopatric ones (implying selection away from the common form) did not support species recognition as a driver of ceratopsian frills and horns.[234] Species recognition does occur, and doubtless features such as stegosaur plates or ceratopsian frills would have served in this role, but recognition does not appear to have been the driver to develop and maintain visual signals—rather to have coopted signals that already existed.

There have been numerous attempts to determine if dinosaurs were sexually dimorphic. Various species have been claimed to be dimorphic in a number of studies, based on widely varying numbers of specimens and different types of evidence or statistical tests (see [118] for *Tyrannosaurus*). Few, if any have stood up to statistical scrutiny (see [119] for a review, although see also [235]), leaving us with the apparent paradox that dimorphism is very common in modern reptiles and birds (and mammals), but apparently rare or even absent in dinosaurs, suggesting little effect of sexual selection. However, two major factors are at play that might make dimorphism hard to detect in Dinosauria.

First, the growth pattern of dinosaurs is typically more reptilian than avian, and this means an extended growth period. With a long time to maturity, and animals capable of reproducing well before somatic maturity,[236] it would be very easy to mistake, say, a large adult female for a mid-size adult male. So even if there was a high level of dimorphism in the species, it might not be apparent without a large sample size (and this may be true of simple body size differences between males and females, but also true where weapons and ornaments may be different sizes, or even present and absent in different sexes).[111] Secondly, in the case of mutual sexual selection, both sexes are ornamented or possess weapons, often of similar sizes (figure 5.3). This has been understudied in living animals until relatively recently, but increasing numbers of examples are known, demonstrating that males competing for females and females competing for males can lead to both sexes having similar ornaments. This is especially the case where high investment in reproduction from males is coupled with a limited number of females that they can mate with. Such a system may explain, for example, the frequent presence of crests or other signaling structures in all known members of a species or clade of dinosaur such as the neoceratopsians (see [34]). Thus the apparent absence of dimorphism does not rule out sexual selection being in operation. Notably, sexually selected features (including those under

FIGURE 5.3 An adult black swan (*Cygnus atratus*), a species which is known to express mutual sexual selection. Increases in the curled feathers on the wings have been shown to be more attractive to males in females and to females in males. They also serve as an expression of social dominance within sexes. Photograph courtesy of Irene Probesting.

mutual sexual selection) tend to be honest signals of fitness (see [237]), and as such are often co-opted to serve as social dominance signals; it may be impossible to separate these drivers for extinct animals. As a result, the term "sociosexual selection" has increasingly been used to refer to this pattern of selection.

Evidence of Visual Signals

Dinosaurs likely had excellent color vision: they would have been tetrachromatic and able to see into the UV spectrum.[238] Given the frequency of visual displays in birds and reptiles (and their often bright colors), it is reasonable to infer that dinosaurs often engaged in similar signaling behaviors. To what extent is unclear, of course; though progress is being made in the elucidation of dinosaur colors and patterns, this data is both limited and subject to caveats. The recent discovery that the bones of some reptiles are sufficiently luminescent to shine through the integument to add to a display opens up this possibility in dinosaurs,[58] though this is currently untested.

Numerous dinosaurs exhibit large bony crests that are likely to have functioned as part of a sociosexual signal (see [34, 239] for reviews), though testing of this hypothesis has, to date, been relatively limited, and most sample sizes for dinosaurs are very small. Multifunctionality is once again a potential issue, although in many cases a biomechanical/functional explanation can be ruled out, leaving some form of signal as the most likely driver for the origin and development of a structure. Animals that used their structures for some mechical function we would expect to converge on one or a limited number of forms, and yet diversity and disparity—even within clades and on closely related species—is the norm, and this is a pattern that is far more common in display features.[233] Active testing for a signal being sociosexually selected has been done based on positive allometry (see Case Study), and here, disproportionate growth late in ontogeny points to a feature that is used only upon reaching sexual maturity.[240]

The ornithischians are the most important group in this context. As far back as Peter Dodson's[241] work on putative sexual dimorphism in *Protoceratops*, there was a firm emphasis on features such as ceratopsian frills being used for display (though comments to this effect predate his work, of course, with Franz Nopcsa writing about dimorphism in dinosaurs as early as 1905[242]). All ceratopsians have some kind of frill, horn, and/or cheek boss combination which typically grow late in ontogeny[172] (figure 5.4), and their sibling taxa in the pachycephalosaurs also have both the bony mass on the head (likely also a weapon—see chapter 7, Combat) and various spikes around the margin as well. All of the lambeosaurine hadrosaurs bear boney cranial crests and show a huge diversity of these, which also appear to grow late in ontogeny.[243] These crests are connected to the nasal passages and would have had a major role in sonic communication (see below). While the saurolophine hadrosaurs generally lack any obvious crest, at least some apparently bore one composed of soft tissues, as seen in a mummified *Edmontosaurus* from the Late Cretaceous of Canada,[8] suggesting that some such form of crest may have been

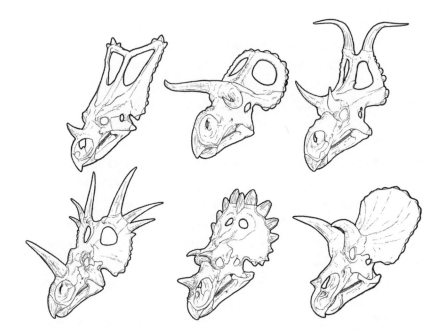

FIGURE 5.4 Ceratopsian heads (top left to right: *Chasmosaurus*, *Nasutoceratops*, *Diabloceratops*; bottom left to right, *Styracosaurus*, *Regaliceratops*, *Triceratops*), showing part of the range of variation of size and shape in the frills, horns, and jugal expansions. This points to a primarily signaling function and not a mechanical one, which would constrain these to a limited number of optimal forms. Artwork by Gabriel Ugueto.

more common still in these animals. The various plates and spikes of stegosaurs ([244]) and ankylosaurians ([245]) were also likely able to function as signals as well as weapons and armor.

Bony cranial crests are also common in theropods, with numerous different hornlets, bosses, and crests known, and there are other putative signals such as the large sail on the backs of the spinosaurids.[34] As large theropods were rare animals in ecosystems, we have relatively few fossils of them, and coupled with their crests also being generally much smaller than those seen in the ornithischians (see below), this has made testing of the signaling hypothesis in theropods impossible to date. Notably, though, younger tyrannosaurs have less well-developed orbital bosses than do large adults (or none), and so the late growth pattern may well hold true here too. Moreover, it is notable that many theropods exhibit bony cranial crests and these are largely replaced by feathers in derived lineages[34, 246]—though in some larger taxa feathers may be reduced or lost[247] while crests are retained. At least some theropods show ontogenetic change in their feathers, with juveniles having smaller and more simple plumes than adults,[248] which on non-flying taxa points to a shift in function through maturity, and again might indicate signaling. A wide variety of feather types are seen in various theropods, including unusual ribbon-like tail plumes,[249] and various theropod feathers show bright colors, some of which are even iridescent.[250] Feathers have some clear advantages over bony crests: they are present across the whole body, can be shed between seasons, can produce more colors and patterns, and can also be manipulated in ways that would not be possible with fixed bony crests.[34] An excellent example is the Cretaceous oviraptorosaurs, many of which have a fan-like arrangement of tail plumes. In one species, at least, the bony part of the tail is dimorphic, such that the bones of the (inferred) male animal are better suited to raising the tail than the female's—implying that the tail fan was under selection for display.[197]

Many of the bony crests of theropods are small, both in absolute size and relative to the size of the bearer, which stands in stark contrast to the larger signals seen in many ornithischians.[34] One possible explanation for this trade-off is the importance of camouflage for carnivorous theropods. As active predators, they would need to be able to conceal themselves from their prey to get close enough to strike, and a large crest would potentially give a warning to their intended targets. Thus counterselection may be in operation, with a trade-off between increasing the prominence of the crest as a sociosexual signal but reducing it for concealment.[34] Such a trade-off could be less important or irrelevant for large herbivores, because they would require long foraging times, and those that may have lived in groups (chapter 4, Group Living) would have been especially exposed over long periods, so it would be difficult for them to hide. Freed from this constraint, signals have the potential to be much larger in herbivores. This might also be true for some theropods if they either were herbivorous (e.g., oviraptorosaurs), primarily scavengers (note that many extant vultures are more brightly colored than hawks or eagles), or operated at night or in cluttered environments. This is perhaps also true for the spinosaurids—which were predators, but may have primarily taken aquatic prey andstruck from above the surface; large signals like the dorsal sail might not have been a strong penalty, even though the crests on the head could have been compromised by exposure to their prey.

Despite their diversity and geological longevity, there is currently limited evidence for sociosexual signals in the sauropodomorphs. It has been suggested that neck length was fundamentally a sexually selected trait,[251] as an explanation for the elongation of necks in various lineages; however, this has been strongly challenged,[252] and adaptations toward feeding are probably the best explanation (see chapter 8, Feeding). This does not necessarily mean that necks were not used as signals, but there is currently no evidence that they were. The clear possible exception is the Cretaceous South American dicraeosaurid *Amargasaurus* (figure 5.5) and its relatives. The dicraeosaurids have unusually short necks for sauropods, but also often have some kind of hyperelongate neural spines. In the case of *Amargasaurus*, these are paired rows of long spines along the cervical vertebrae that may have been linked together to form a kind of muscled "sail."[253] There are no obvious functional benefits to this arrangement, but in an animal that is (relative to other sauropods) small and short-necked, such a feature would serve well as a visual signal.

The large frills of ceratopsians could have served as an aposematic signal to warn off rivals, or in particular, predators, especially in combination with their often large horns.[34] An animal could "flash" these

FIGURE 5.5 A life reconstruction of the small dicraeosaurid sauropod *Amargasaurus*, with its unusual paired rows of elongate spines that extend from the cervical vertebrae. Here the spines are joined together to form a sail-like structure as based on [253]. Artwork by Gabriel Ugueto.

by lowering its head to display the frill. One possible counterargument is that if predators were such a threat to adult animals with fully developed frills, then it might be expected that frill morphology would converge through Mullerian mimicry, such that an encounter with one species would serve also to protect others, providing mutually beneficial defense. However, as noted by Andrew Knapp and colleagues,[234] even contemporaneous and sympatric ceratopsians that would have faced the same predators have different frill and horn shapes and combinations, and this suggests that aposematic signaling would likely be a secondary function, if it existed at all.

Camouflage may be considered something of an "anti-signal," by which animals attempt to hide themselves from predators (figure 5.6). It functions by effectively reducing the signal of the animal, or by decreasing the signal-to-noise ratio, thereby making the animal hard to detect.[254] The best example of camouflage known in dinosaurs so far is the small ceratopsian *Psittacosaurus* from the Early Cretaceous of China. An exceptionally well-preserved specimen of this genus preserves evidence of its skin patterns as well as quill-like extensions on the proximal half of the tail.[255] A highly detailed and accurate 3D model of the animal was made and tested under varying light conditions associated with the known local ecosystem to show that the colors and patterns likely functioned well to camouflage the animal in a wooded environment.[255] Notably, the animal showed countershading, with a lighter underside to reduce the influence of shadow there and conceal the outline of the animal (figure 5.7). This patterning is extremely common in living species and likely was widely used by many dinosaurs.

The nodosaurid *Borealopelta* similarly shows countershading across the animal as a whole, yet also has brighter colors on the largest of the lateral spines.[245] This points to the possibility of counterselection and the need to balance differing signals with differing selective pressures (as in the above example of theropod cranial crests). Without sufficient positive signals to attract mates or to put off rivals, there will be selection for brighter colors or larger areas of color, but if this is too great, the animal may be too conspicuous and vulnerable to predation.

FIGURE 5.6 Camouflage can be considered an anti-signal in trying to avoid detection. Left, three tawny frogmouths (*Podargus strigoides*) are well hidden as a result of their color, pattern, and posture. Photograph courtesy of Michael Buckland. Such effects are context dependent, however—the snowy owl below (*Bubo scandiacus*) is not well hidden when there is no snow behind it. Photograph by the author.

FIGURE 5.7 Life restoration of the small ceratopsian *Psittacosaurus*, based on[255]. The patterns of a single, extremely well-preserved individual are known in detail and show that this animal was likely well camouflaged in the forest environment it inhabited. Artwork by Gabriel Ugueto.

Evidence of Auditory Signals

In their range of auditory signals, dinosaurs were likely limited to predominantly closed-mouth vocalizations,[256] although this would still have provided calls such as the roars, hisses, and others made by crocodilians and indeed other reptiles,[257] while precluding the more complex calls of derived modern birds with a syrinx. Many, and perhaps all, dinosaurs were at least capable of producing some sounds, and would have done so to communicate.

At least some members of Dinosauria were well equipped to make large and loud noises that doubtless would have been a major part of their signaling. Most obvious are many of the crested lambeosaurine hadrosaurs, which have large and elaborate cranial ornaments that are connected to the nasal passages and would have served as excellent echo or amplification chambers[258] (figure 5.8). While not yet studied in detail, these show a fundamentally similar pattern to that observed of the sociosexually selected crests noted above: absent in young juveniles, rapidly developing late in ontogeny, and apparently

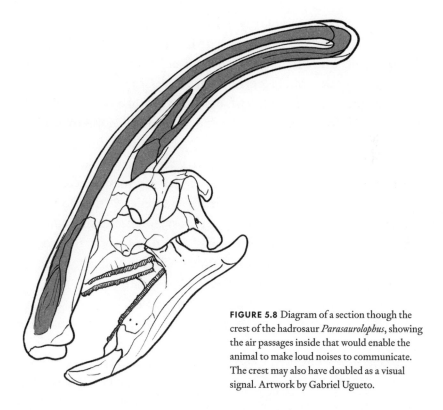

FIGURE 5.8 Diagram of a section though the crest of the hadrosaur *Parasaurolophus*, showing the air passages inside that would enable the animal to make loud noises to communicate. The crest may also have doubled as a visual signal. Artwork by Gabriel Ugueto.

having evolved near randomly within the clade. Moreover, some show both that the development of these nasal passages is not congruent with any change in olfactory ability and that the ear was well-suited to hearing the kinds of low-frequency sounds that these crests would produce.[69] Coupled with an increase in brain size that may point to an increase in social behaviors, this is a comprehensive argument for these crests being used to generate or modify calls to communicate.[69] These crests may have doubled as a visual signal (as multiple signals from multiple different sources can be important—see [259]), but the vocalization component was probably the dominant force in their evolution.

Although the most prominent example, hadrosaurs are not the only likely vocalizer. Many ankylosaurs have enormously complex nasal passages that can even loop inside the skull and have been hypothesized to have had an acoustic function.[260] Coupling these observations with work in other clades and with data on pitch-attunement of the ears makes a credible case for vocalizations being important in this group ([57], and see chapter 3, The Basis of Dinosaur Behaviors). Recently this evidence has been greatly enhanced by the discovery of a large, kinetic ossified larynx from a Late Cretaceous Mongolian ankylosaur that is avian-like (though not a birdlike syrinx); this was likely suitable for modifying noises and may have been able to produce some birdlike sounds.[261] An interesting potential conflict here lies in a provisional study that suggested that ankylosaurs had a reduced tympanum.[262] Together with the discovery of a fossilized cartilaginous voice box from a Cretaceous bird in Antarctica,[263] the Mongolian find at least provides the knowledge that such softer tissues could be preserved for more non-avian dinosaurs, and so a larynx from other dinosaurs (ossified or not) may yet be recovered and give a deeper insight into their ability to vocalize.

Sounds need not have been generated vocally, and animals have evolved various other forms of noise generation, such as the beak clacks of storks and other birds (see [239]) and the knee clicks of eland (*Taurotragus*[259]). One hypothesis in dinosaurs is that the greatly elongate whiplike tail of diplodocid sauropods could be flexed with such speed as to create a supersonic crack, with the potential to signal to others, be it members of a group, potential mates, or as a warning to potential predators.[264] In this case it is at least possible to make a biomechanical analysis of the skeleton to show whether this could be achieved—and one by Simone Conti and colleagues[265] suggested that it is not. Other options (such as a simple stamp of the foot) would have been available to many dinosaurs, but almost impossible to assay.

Evidence of Other Signals

There is currently relatively little evidence for other signals in dinosaurs, though undoubtedly they would have employed touch, taste, smell, and other senses. The use of olfaction can be at least partly inferred from brain structures, and we know that some dinosaurs had limited olfactory capacity, while it was well developed in some theropods (more so in carnivorous than herbivorous taxa[59]). Here, it would be difficult to separate out the role of strong olfactory capacity in actual deliberate communication, since olfaction has a number of other roles, such as in finding food and detecting threats. However, given how commonly scent is used for both marking territories and communicating reproductive state in many tetrapods, including many reptiles[266] and birds,[267] it is certain that this would have been employed by numerous dinosaurs. One possible candidate is early ceratopsian *Psittacosaurus*, which shows apparent swellings at the cloaca of an extremely well-preserved specimen that might be homologous with musk glands seen in modern crocodilians.[268] If so, this could demonstrate the potential for dinosaurs to produce chemical signatures for olfactory communication.

Dinosaurs must have had some form of mechanoreceptors, and touch must have been an important sense that doubtless would have been used in various forms of communication (beyond more obviously "tactile" signals such as an impact from a horn or club—see chapter 7, Combat). It has been hypothesized that tyrannosaurs had unusually sensitive snouts, which could have been important in tactile signals and courtship,[130] although this interpretation was based on the neurovascular foramina in the jaws, and these may not be any more common in tyrannosaurs than in other large theropods[63] (figure 5.9).

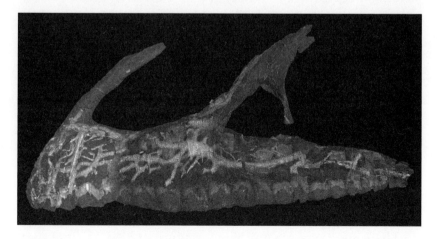

FIGURE 5.9 CT scan of the maxilla of the Early Cretaceous carcharodontosaur *Neovenator* showing the tracks and foramina though the jaw. This has been suggested to indicate high sensitivity in the muzzles of some theropods. Photograph courtesy of Chris Barker.

Also, given the very high number of craniofacial injuries seen in large tyrannosaurs,[37] a highly sensitive face could have been something of a problem if combat was very common, and so quite how sensitive these areas were is open to question.

One last option is the manipulation of the environment as a signal. A number of sites show multiple sets of parallel marks scraped into the substrate by the feet of large theropods. It has been convincingly argued that these are the result of courtship rituals between animals, as they are a surprisingly close match for traces produced by a number of seabird species that carry out very similar displays.[35] In this case, the evidence comes directly from ichnology rather than from the dinosaur body fossils themselves, and is most unusual in this regard. It also raises the possibility that dinosaurs may have engaged in other such courtship behaviors that would not have left clear traces, such as gift-giving or bower construction.

Summary

Although the exact mechanism of the delivery of signals and their reception is difficult to determine, it is clear that numerous dinosaurs were communicating in a variety of ways. Moreover, in a large number of species and clades we can make a strong case for sociosexual signaling, based on allometry and the development of exaggerated structures that appear only late in ontogeny and do not have clear biomechanical functions. The details may be missing (e.g., degrees of dimorphism, colors of signals, how they were displayed), but this pattern points to the importance of communicating signals within species, and gives a clear indication of the potential for complex signals and behaviors in many dinosaur groups. There are also major implications here for the potential of social behaviors through dominance signals, and for reproductive behaviors through high mutual parental investment in offspring.

There is a clear pathway here for future studies. As more specimens are found (especially those of younger individuals), we will have more opportunity to look for patterns of positive allometry and sexual dimorphism, and to test the functionality of various bony and soft-tissue crests. Similarly, work identifying colors will only generate more data to illuminate possible changes in ontogeny or differences between the sexes, and will allow the testing of further hypotheses about the function and evolution of these features. Continuing work on the senses of dinosaurs will also help frame hypotheses about whether or not certain signals could be preceived (especially sounds), and we have yet to try to begin to try to calculate how far sounds of certain pitch or volume, likely made by some dinosaurs, might travel in various environments. Integrating these approaches could well provide a more solid platform for hypothesizing how these animals may have communicated and the selective pressures that shaped their evolution.

The focus in this chapter has very much been on sociosexual signals, since these are major aspects of dinosaur behavior, there is a clear method to test their function, and there is ample data to work on. However, other aspects of communication should not be overlooked going forward. Doubtless various dinosaurs engaged in more complex communication dynamics such as brood parasitism, aposematic mimicry, or olfactory camouflage of nests. These might be impossible to test without the appropriate fossils, but should not be overlooked, and future focus should not be solely on large display features.

CASE STUDY: **Multiple signals in *Protoceratops***

Protoceratops is a small neoceratopsian from the Late Cretaceous of Mongolia, and one of the best-known dinosaurs. Dozens of complete skeletons have been preserved,[241] and following recent new discoveries, this includes very young animals[269] that may have still been in the nest when they were buried. In all, we have an exceptional representation of animals, from young juveniles through to large adults, from a single geological formation (figure 5.10).

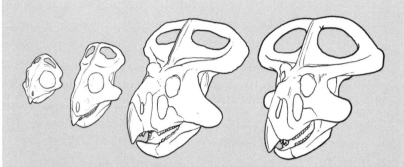

FIGURE 5.10 A series of skulls of the Late Cretaceous *Protoceratops* showing the change through ontogeny and the rapid expansion of the frill during growth (left to right: hatchling or very young juvenile; juvenile; subadult; adult). Series based on [271]. Artwork by Gabriel Ugueto.

The large frill at the back of the head has long been considered likely to be some kind of signaling structure, sexually selected or otherwise, and also thought to vary between individuals, representing both male and female morphs. This was first explored by Peter Dodson,[241] who advocated a clear split between generally smaller females with a smaller frill, and larger male animals with a proportionately larger one. However, subsequent analyses using larger datasets and more detailed measurements have failed to recover this split.[270] It remains possible that these animals were dimorphic, but the dataset is still insufficient to demonstrate this. In either case, the frill would appear to be a good candidate for a signaling structure of some kind, and in the absence of other clear mechanical functions, this hypothesis was tested.

My colleagues and I obtained measurements of *Protoceratops* skulls of multiple different ages, from putative hatchlings up to large, and presumably old, adults.[271] Notably, all *Protoceratops andrewsi* specimens are from a limited geographical area and temporal distribution, meaning that these can be treated as being as close to a natural population in the dinosaurian fossil record as it is normally possible to be.[272] We found that the parietal frill shows strong positive allometry in both length and width, demonstrating that the frill grew disproportionately larger as an animal increased in size, while it was very small in animals below the size of likely reproductive maturity, and therefore could be linked to reproduction. In short, this feature was likely under sociosexual selection. This was further supported by Devin O'Brien and colleagues,[50] who showed that this rate of growth specifically matched that seen in numerous extant vertebrate taxa whose various display features are known to be sexually selected.

In addition, work using 3D scans and models of numerous *Protoceratops*[272] demonstrated that the frill was effectively a separate unit and grew and changed far more rapidly than other parts of the skull during ontogeny. Although fewer specimens were complete enough to measure in this study, it was far more in-depth than earlier ones that had relied on a limited number of simple lineage measurements. The frill may not, though, be the only signaling structure in *Protoceratops*. My colleagues and I also noted that the large jugal expansions on the cheeks of the animals appear to grow much faster than the rest of the skull as well, and other ceratopsians also have large cheek bosses and even spikes.[271] Furthermore, the middle part of the tail of *Protoceratops* has unusually long neural spines, giving it a sail-like appearance, and these expansions do not appear in the youngest animals. Previous work[273] had noted that the tails of these animals were structured so that they were well suited to being elevated in the air, and thus might support the idea that the tail functioned as an additional signaling structure.

However, uncertainty remains. In the case of the tail, no work has been done to measure and assess the growth rate of the caudal neural spines to look at their allometry, and data to test this is harder to obtain, as complete tails are rare. Similarly, the study by Andrew Knapp and colleagues[272] failed to find that the jugals grew at an unusual rate compared to the rest of the skull, suggesting that they might not be unusually large in adults.

Nonetheless, *Protoceratops* remains an excellent example of assessing if a feature is likely operating under sexual selection. Demonstrating strong positive allometry across a wide range of specimens of different ages, and lacking any clear mechanical explanation for the frill's appearance, shows that this is a feature in use only by adults, and that the largest individuals support disproportionately large signals—as would be expected for healthy, dominant animals.

CHAPTER 6

Reproduction

The selective pressure of reproduction is of course one of the biggest in nature and is central to natural selection through inheritance and propagation. However, sexual selection will play an arguably bigger role in shaping the biology and behavior of a species. Sexual selection may be considered "any selection that arises from fitness differences associated with nonrandom success in the competition for access to gametes for fertilization,"[231] while others would go further to suggest any selective forces that act on reproductive success. In either case, this is clearly a powerful driver of evolution. Here we address the behavioral impact on dinosaurs from the act of mating to nest building and parental care. (Courtship is considered as part of chapter 5 on Signaling, although various species might also have engaged in fights for mates; see chapter 7, Combat.)

There is huge variety in the reproductive systems of modern tetrapods that provide potential models for dinosaur behavior. In a number of cases the ecological and behavioral effects of different reproductive systems can be linked together (see [274]), though to date this approach has been little applied to dinosaurs. Starting with the most fundamental aspect of reproduction, males typically produce low-cost sperm, and females produce high-cost eggs; and (in vertebrates with internal fertilization) females are guaranteed maternity of any given fertilized eggs, while even males that have successfully mated with multiple females may not have any offspring. Those differences in investment and differences in parental certainty clearly have the potential to lead to different selective pressures on males and females. Coupled with ecological and environmental differences (e.g., density of predators, or available food), these selection pressures can then lead to very different strategies in reproduction between the

FIGURE 6.1 Life reconstruction of the growth of the "prosauropod" *Mussaurus* from egg to adult. Unusually, these animals appear to have shifted their fundamental locomotory system as they grew, changing from quadrupeds to bipeds. This must have changed their general ecology significantly as they grew. Artwork by Gabriel Ugueto.

two sexes, and even within sexes (see [275]). However, determining if females were polyandrous or males defended harems of females, or the extent of any extra-pair copulations, is essentially impossible. Our interpretation of such behaviors can only be inferred from our understanding of the drivers of reproductive systems in extant animals, coupled with some critical ecological and behavioral data (and here communication may be especially important). Even so, it must not be overlooked that some dinosaurs might have been unique in their reproduction and development (figure 6.1).

Much of this variation in mate choice and reproductive strategies is linked to investment in offspring, both pre- and post-hatching parental care, and, by extension, the classic split between "r" and "K" selected strategies. Under r selection, species typically produce numerous offspring with relatively little investment into each one, whereas K selected species have fewer offspring but with much higher investment in them.[276] These are, of course, extremes, with a whole gradient of differing degrees of numbers and effort between the two; the r/K binary oversimplifies an enormous diversity of strategies for commitment of resources to different aspects of reproductive output and offspring survival. Dinosaurs do, though, show are least some degree of both strategies (and therefore sit somewhere between the two), typically having had both large numbers of eggs *and* post-hatching parental care (see below).

Mating

Although dinosaurs display numerous features that are likely under sociosexual selection (see chapter 5, Signaling), exactly how they used these to display as part of courtship, or what else they may have done, is effectively unknown. As noted by Timothy Isles,[239] there is a very wide range of courtship behaviors in both extant reptiles and birds which could provide potential models for dinosaurs. However, faced with so much variation, and with no clear patterns linked to, e.g., size or ecology, it may be fruitless hypothesizing to pick any one type over another in dinosaurs, based on the current available data. The one exception is the apparent courtship grooves described by Martin Lockley and colleagues,[35] which likely represent some kind of ritualized action (see chapter 5, Signaling). Timing would be key to ensure that hatching would ultimately be aligned with optimal conditions, and we at least expect that animals in high latitudes were very seasonal breeders, with year-round mating more likely at low latitudes with more consistent conditions through the year. Starting with Stephen Emlen and Lewis Oring,[277] we have built strongly in our understanding of how aspects of biology such as sex ratios or environmental conditions can promote, say, polyandry vs. polygyny, or sexual dimorphism. The data required to investigate such factors in dinosaurs is still essentially lacking, however, and so they can only be hinted at in a few cases.

The actual mechanics of mating in dinosaurs has been little explored, though the obvious issue is getting the relevant parts of males and females together. Again, a variety of postures are possible based on those of extant animals, though most likely is perhaps a reptilian "semi-lateral" pose, with the male half on his side to reach partly under a female (as practiced by many squamates and crocodilians[239]). However, there are obvious constraints—those animals with sails

FIGURE 6.2 Hypothesized mating posture for *Stegosaurus* based on [239]. The plates and body shape combine to make coupling potentially difficult, and even more so for giant sauropods. Art by Gabriel Ugueto.

(spinosaurids) or with various forms of armor (stegosaurs) would struggle to achieve this. Then there are issues, like the extremely wide, rigid, and low-slung bodies of anklyosaurs and the sheer size of most sauropods, that would make coupling difficult (figure 6.2). Determining quite how these animals might have joined is difficult, though at least biomechanical analysis allows us to test, for example, whether or not heavy dinosaurs could rear up and whether a female could support the partial weight of a male (see [278] on rearing in sauropods), while the suggestions that damage to dinosaur tails is a result of stress from mating with heavy partners require further study. The simple conclusion to these questions is that males possessed some kind of phallus or intromittent organ that would allow them to reach the cloaca of a female. This is common in both reptiles and paleognath birds,[279] and so was likely present in dinosaurs, though the exact shape, and in particular the length, would clearly vary. Nothing is known of such an organ currently, though the recent description of an external opening of the cloaca for a specimen of the early ceratopsian *Psittacosaurus*,[268] showing a dorsoventral long axis similar to that of crocodilians, gives hope that one may be preserved in the fossil record.

Nest Construction

There are numerous examples of nests of dinosaur eggs preserved that represent a great many different lineages of dinosaurs, though the recent discovery that some (and perhaps many, especially in the early groups) had soft-shelled eggs[280] likely explains why these have yet to be found for various clades. Actual laying could be very disorganized, with large numbers of eggs simply dropped irregularly into the nest, or highly structured, with eggs carefully arranged into rings or rows, and sometimes in several layers[281] (figure 6.3).

There is an essential split between "closed" and "open" nests. All modern crocodilians bury their eggs in closed nests, generally with vegetation to provide humidity and heat from decay to incubate the relatively soft-shelled eggs (figure 6.4), while (nearly) all birds have open nests where the body heat of the brooding parent incubates a harder-shelled egg. Softer, or at least thin-shelled, eggs such as those of various sauropodomorphs[282] and early-branching ceratopsians[280] must have been buried to keep them from drying out,[283] although

FIGURE 6.3 Camera trap photograph of a female mugger crocodile (*Crocodylus palustris*) excavating her nest and lifting out an unhatched egg. Adult crocodilians will help hatchlings free themselves of the nest. Photograph courtesy of Phoebe Griffith.

FIGURE 6.4 Nest of large theropod eggs from the Late Cretaceous of China (possibly of the giant oviraptorosaur *Gigantoraptor*). The circular pattern with a space in the center and eggs laid in pairs is seen in numerous birdlike lineages of feathered theropods. Photograph by the author.

thicker-shelled eggs could have been either brooded by parents or buried, or a combination of the two.[284]

Nests would have been excavated from the local substrate, most likely using the hands or feet. Indeed, it has been suggested that the unusually large claws on some sauropod feet functioned primarily to excavate nests through scratch digging.[157] Many dinosaur nests are basic circular hollows in the ground where the substrate was simply dug up and piled around the rim (see Case Study). One notable example is the numerous nests of titanosaurian sauropods from the Late Cretaceous of Argentina.[285] Thousands of eggs are known across at least four layers, suggesting site fidelity by the species over an extended period. The local paleosoils were of clay, but the excavated nests are infilled with silt from flooding, so the eggs were not buried in soils, though other titanosaur nests preserve traces of vegetation suggesting that this was used to cover the eggs.[42] Nests were in a high density, suggesting gregarious nesting habits, and were irregularly spaced in some cases, with nests on top of each other.[285] Such an arrangement was likely common for many dinosaur species.

Others would have had at least partially open nests. Nests from birdlike theropods are known that contain elongated eggs laid vertically, and these were only partially buried in sediments, with the upper two-thirds of the egg exposed.[286] These would be incubated either in decomposing vegetation or by body heat from the parents (see below). Elongated eggs would have been resistant to compression in the long axis, so would have been well able to withstand any pressures from this unusual arrangement.[281] One interesting recent discovery is that of pigmented eggshells for some derived theropod dinosaurs that show both major colors (red-brown and blue-green) that produce all the variety of modern bird eggs.[287] The presence of shell pigments suggests that the eggs were partially or fully exposed (not covered by soil or vegetation), and evidence for a cuticle coat that would reduce water loss in these eggs is known from fossil eggs from arid environments, further supporting the idea that these were not fully buried, and suggesting a mixed incubation mode of animals sitting on partially buried eggs.[288]

Egg Laying

Dinosaurs had two functional ovaries, and some derived theropods apparently produced only two eggs at a time (one in each ovary), meaning that eggs were laid in pairs, but also that it could take many days to lay a complete clutch of numerous eggs.[289] Nest structure, egg laying, and brooding are likely linked together, given the risk of damaging eggs while burying them. In short, animals like sauropods laying large numbers of disorganized eggs likely laid these all in one sitting and then buried them. To lay several eggs every few days would mean either uncovering and reburying the nest, repeatedly exposing them between deposits, or adding ever more layers of sediment, and either would risk damaging the eggs. However, for those species with partially buried or fully exposed eggs that the parent incubated, a more birdlike pattern was likely: a pair of eggs laid down at a time and the parent then incubating and protecting the eggs while the laying cycle for the clutch was completed.

Clutch size is an important component of reproduction, as invest-ment here in eggs is linked to levels of parental care, and we would expect this balance to be optimized for the species to maximize repro-ductive output.[290] Egg size is also important in terms of investment in offspring, although in dinosaurs there are critical physical limita-tions and differences between small and large dinosaurs because of the mechanical limits on egg size. Eggs that are too large would require such a thick shell to support their own weight that the embryo would suffocate due to limited gas exchange, so although small dinosaurs could lay proportionally large eggs, large taxa could not. The largest known eggs of dinosaurs are actually smaller than those of the largest prehistoric birds, which could reach 14 L in volume, while those of sauropods, for example, average only 1 L[281] (figure 6.5).

Although numerous nests are known from dinosaurs showing widely variable numbers of eggs (e.g., the oviraptorosaur *Citipati* with 15 eggs;[42] titanosaurian sauropods with up to 40 eggs[285]), it is not possible to tell if the contents of a single nest represent the re-productive output of an individual animal. Dinosaurs may well have had communal nests, or nests dominated by single individuals to which others contributed (as seen, for example, in ostriches, *Struthio camelus*[291]), and those that did not directly brood their own eggs might have laid multiple nests of eggs in quick succession. In short, a fossil nest of eggs does not necessarily represent the reproductive output of a single female, which could be much higher or much lower.

Clutch size may be linked to factors such as female body mass, parental investment, nestling period, adult mortality, and others,[292] only some of which can potentially determined, or even estimated, for dinosaurs. However, the correlations are not always clear: for exam-ple, previous attempts to correlate clutch size with female body size to infer the degree of parental care have proven problematic because of the differences in reproductive systems and other factors that affect clutch size, and these may be difficult to control for (see [293]). Even so, some factors are known to have strong effects on clutch size, including body size and latitude,[294] and so this remains an area that could be explored further and potentially yield useful information about gross patterns of dinosaur investment in and care of offspring.

FIGURE 6.5 The embryo of a titanosaur sauropod preserved inside the egg. Embryos allow the identification of eggs far more accurately than the shells alone and provide information on the developmental rate of animals prior to hatching and the potential reproductive output of single individuals, matching the volume of eggs in a clutch to adult size. Photographs courtesy of Martin Kundrát.

Incubation and Brooding

Work on brooding behaviors and incubation in general has tended to focus on the feathered dinosaurs closest to birds (oviraptorosaurs, troodontids, and dromaeosaurs), both because of the question of the transition to birdlike behaviors, and also because the specimens that best demonstrate brooding behavior come from this clade. That is in itself perhaps unsurprising: if most dinosaurs simply buried their eggs in soil or vegetation (even if they defended or manipulated these nests), we are unlikely to find the adults buried alongside them. However, we now have a number of specimens of adults preserved sitting on eggs (especially oviraptorosaurs—see [43, 295]), so there is no question that these particular animals did sit on their nests at least part of the time and, as noted above, likely provided at least some of the necessary warmth to the eggs. The evidence—feathers for insulation, the structure of the nest with unburied eggs, eggs with colors and protective coats, and eggs laid over time in pairs—all intersects to support this interpretation. The partial burial of some eggs in nests in an upright position, with the sediment holding them in place, suggests that they would not have been turned by the parents, as bird eggs usually are,[281] although this leaves open the possibility that in open nests, the eggs may have been laid "flat."

Not all feathered dinosaurs would have brooded in this way, however. Animals like large tyrannosaurs may have had feathers, but while these would have provided insulation for the individual, they were small and filamentous, not pennaceous, and would have been of limited help in warming eggs—not to mention the obvious issue of how a multi-ton animal might have safely sat on a nest. It is not certain exactly what feathers (either type, extent, or size) were present on various parts of the body of several feathered lineages (e.g., ornithomimosaurs, alvarezsaurs, therizinosaurs), and so, in the absence of known nests, the brooding patterns of these are unknown for now. At least some of the most birdlike were probably capable climbers (see chapter 3, The Basis of Dinosaur Behavior), opening up the possibility that these animals were sufficiently arboreal to have nested in trees. However, such nests would be extraordinarily unlikely to preserve, and none are currently known. Even if they were found, they might be hard to distinguish from those of contemporary avialians.

As noted above, other dinosaurs likely simply buried their eggs. At least some titanosaurs used local geothermal heat from volcanic soils to incubate their eggs, a method that is also used by a number of extant megapodiid birds today,[296] so not all needed decaying vegetation to provide heat. Even those eggs that were buried might have needed parental attention, both for general protection from potential predators and to help regulate temperature by clearing away or adding cover to the nest.

The incubation periods of dinosaurian embryos, as determined by counting growth lines on their teeth, show a generally reptile-grade level of development,[44] and embryos from some soft-shelled eggs suggest very long incubation times, potentially many months.[280] This has important implications for the question of pre-hatching parental care. While such care is normal for archosaurs, and would be the ancestral condition for dinosaurs, long incubation times would change the dynamics for parents. Adults could potentially have guarded the nests throughout incubation, but this would require very extended stay in the area and would potentially make foraging difficult for large animals. A whole group of sauropods could not likely stay in one place for that long without running out of food (though this doesn't rule out a strategy like an extended reduction in intake).

Moreover, reproductive output and how investment was delivered to the next generation would greatly affect dinosaur behavior. If sauropods were, for example, laying only limited numbers of small eggs, this would imply considerable parental investment through pre- and post-hatching parental care, but if they were laying multiple clutches of numerous eggs each breeding season, these greater numbers of offspring (and more effort going into simply producing them) might mean that care was limited, or even nonexistent, after construction of the nest. Alternatively, animals might leave after egg laying only to return to find their nest close to hatching time and protect their offspring then. In either case, however, this implies potential extended care and investment from at least one parent, and possibly both. Division of such efforts would likely link to selection pressures on mate quality and, by extension, mate choice, and link back to questions over dimorphism, mutual sexual selection (see chapter 5, Signaling), and cooperation (see chapter 4, Group Living).

Post-hatching

Most, though not all, crocodilians show post-hatching parental care,[105] and post-hatching care is near universal in birds (predominantly by both parents[297]), so some form of care would likely be the ancestral condition and default hypothesis for most dinosaurs. Such investment can take a number of forms, from helping the young to leave the nest, to guarding the offspring and defending them from predators, to helping keep them warm or cool, to feeding them (or at least leading them to food) and teaching them to forage successfully. Care is highly varied in birds, from limited or even no care for pre-hatched offspring (beyond construction of the nest), through to male-only, female-only, biparental, and cooperative group care[298] (figure 6.6). Once more, given the variety of dinosaur ecology, it is likely that degrees of care were highly variable, given the different pressures that can lead to differing levels of male investment.[299]

Interestingly, there is evidence for male care in feathered theropod dinosaurs. Brooding specimens found on nests of eggs have been diagnosed as males based on the absence of medullary bone, a store of calcium for eggshells which is laid down by females during the breeding season (see [300]). In birds, at least, females typically provide more care pre-hatching, but post-hatching care is split equally between the sexes,[301] so male-only brooding would probably be rare, and male brooding would be more unusual than male post-hatching care. However, in addition to some of the inconsistencies in associating the presence of medullary bone with an animal being female,[302] its absence could indicate a number of different scenarios. The brooding animal might indeed be male, exhibiting male-only care, or there could be biparental care but the male happened to be on the nest at the time of death; it could be part of a group, or even a young or misdiagnosed female. Such uncertainty could potentially be resolved

FIGURE 6.6 Above, an adult female gharial (*Gavialis gangeticus*) with a collection of young juveniles. Post-hatching parental care is being offered here in the form of protection. Below, an adult kookaburra (*Dacelo novaeguineae*) shows extended parental care by feeding a large and fledged chick. Photographs courtesy of Phoebe Griffith and Neralie Thorp, respectively.

with additional discoveries: for example, multiple exclusively male specimens found on nests, or equal numbers of males and females, would strongly suggest different care regimes. Further support for different regimes of care could come from studies of the general biology of the species, such as increased brain sizes and certain ecological lifestyles correlating with increased parental investment[187]—and, notably, increases in extended parenting are linked with increased cognition (in corvids, at least[303]).

Evidence of post-hatching care primarily comes from finds of single adult animals together with multiple juveniles (e.g., *Psittacosaurus* represented by an adult with some 34 juveniles,[304] and an adult thescelosaur, *Oryctodromeus*, in a burrow with two juveniles[305]). These examples are limited in number, and obviously the exact relationships of the animals preserved are unknown, so while it is reasonable to infer that it is a parent with offspring, it is at least possible that the adult was one of a pair with the other parent missing, or was a related helper (i.e., a sibling from an earlier clutch), or even that this was just a chance association. There are also multiple examples of mixed-age groups of dinosaurs (see chapter 4, Group Living), which may potentially represent cooperative breeding, though more likely they are simply aggregations of juveniles and adults (although these might include multiple sets of parents and offspring). The other major set of evidence of post-hatching care comes from large juveniles in nests which were evidently fed there (see Case Study), and so providestrong support for extended care in some hadrosaurs, at least.

There are also large numbers of groups known of just juveniles, apparently separate from adults, including some of small sizes that were very young. This leads to the suggestion that juvenile-only aggregations were a common feature of young dinosasurs.[306] Juveniles do suffer special penalties—they are especially vulnerable to predation[88]—and so crèche-like groups would likely form among nestmates or animals from the same cohort in the absence of parental care (see chapter 4, Living in Groups). Various theropods are known from such aggregations of juveniles,[306] and in the case of sauropods we see tracks of young (and small) animals in groups that are apparently close to their hatching site, not traveling like the adults in the region.[307] This implies that the juveniles were not going out to forage and so were perhaps looked after in some way, though they would need far less food than the adults and so may have been living apart from them. In the case of ornithischians, a juvenile cluster of very young *Protoceratops* is known that were originally described as nestlings,[269] but have also been suggested to represent an independent group.[214] Of greater significance is a group of six *Psittacosaurus*,[31] all juveniles, but of two distinct ages/sizes, that are around 1.5 years apart. These are suggested to represent multiple clutches, which is likely, though there is no direct evidence that they were related. Certainly it appears that larger juvenile *Psittacosaurus* may have lived together after the end of parental care, as seen in this specimen.

Collectively, these finds represent considerable variation, with many intriguing, but singular, data points for various species suggesting male parental brooding, post-hatching parental care, or abandonment of juveniles. However, as with the case of potentially brooding males, there is a strong need for more specimens of the species known from juveniles, with *Psittacosaurus* the only animal potentially having demonstrated some change in behaviour across ontogeny. Nests with adults, and multiple specimens of adults with juveniles would allow us to assay the possible patterns of care and while this is currently lacking, it has future potential.

Summary

From the discovery of the first dinosaur eggs in the 1930s, our knowledge of dinosaur reproduction has advanced considerably, and we have an excellent understanding of some aspects of their reproductive behavior. Evidence of nest building, egg laying, incubation, and post-hatching parental care is known, with both an excellent fossil record and scientific interest providing a strong focus on the birdline theropods.

Other areas remain little known or poorly understood. We have little evidence for specific courtship actions or mating behaviors, though with promise for greater understanding of both in the near future. Similarly, our growing understanding of the links in extant animals between parental care, mating strategies, mate choice, and investment in offspring is likely to yield applicable data and models for interpreting dinosaur behaviors. While it is difficult to determine how these animals were acting, the data on sociosexual signaling and dominance with evidence of pre- and post-hatching parental care combined with existing data on body size and local climates could be sufficient to produce much better hypotheses of dinosaur reproductive behavior. That said, uncertainties over key aspects such as clutch sizes and reproductive effort will likely prove a major issue in these endeavours.

Novel discoveries continue to offer scope for new hypotheses. Brood parasitism is known to be a driver of increasingly complex eggshell patterns in birds, for example, opening up the possibility that this was also an issue in dinosaurs. With the recent evidence that derived theropod eggs were colored, this has the potential to enable testing of this hypothesis in those species with exposed nests, and to shed new light on future assessments of egg colors and patterns. Similalry, integration of sufficient data from tooth isotopes and medullary bone with parents on nests, coupled with incubation times, could reveal if breeding was seasonal or year-round.

CASE STUDY: Pre- and post-hatching parental care in *Maiasaura*

Compared to dinosaur body fossils, dinosaur eggs remain rare, and young juvenile dinosaurs rarer still. Comments on this rarity go back a long way in the dinosaurian literature, and it was long argued that either young dinosaurs were rare simply because they had few offspring, or that eggs and nests were limited to uplands and so not often preserved (see [308]). However, while subsequent decades of discoveries have now revealed numerous eggs, nests, embryos, and even brooding parents on nests (see [281]), these are typically only single finds, which provide limited amounts of data and insight into the extent of parental care and the behaviors being exhibited by putative parents. In contrast to this general pattern is the site in Montana commonly called "Egg Mountain," which provides a wealth of data and allows for an in-depth look at how at least one dinosaur—the Late Cretaceous hadrosaur *Maiasaura*—attended to its offspring.

The name *Maiasaura* (appropriately enough meaning "good mother reptile") was formally given to the skull of an adult hadrosaur, but this was done because of its near association with a far more important find.[284] That was the remains of a total of 15 skeletons of juvenile dinosaurs, each around 1 m long and found in direct association with a nest. The nest was a low mound, several meters across and over 1.5 m high, with a large and bowl-shaped depression in the middle that was nearly 2 m in diameter. Eleven of the skeletons lay inside the depression, with the others scattered nearby. Scattered inside the nest were the fragments of broken eggshells.[284] Importantly, the young animals showed evidence of wear on their tooth batteries and were far too large to have recently hatched from the eggs. The conclusion reached was that these animals had hatched long before but had stayed in the nest while growing and were fed in situ by one or both parents[284] (figure 6.7). Precocial animals would surely have left the nest to forage alone, and returning to an open nest without protection would have offered little defense from predators, so the only clear rationale for staying in the nest would be if food was brought to them. Although

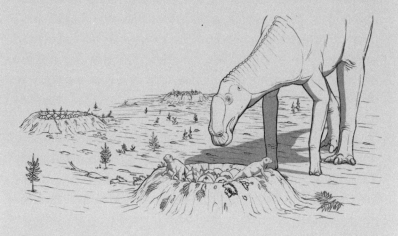

FIGURE 6.7 Life reconstruction of *Maiasaura* nests with an adult tending to offspring. Artwork by Gabriel Ugueto.

initially it was suggested that the long bones and even tarsals of these juveniles were well ossified,[284] later work showed that these had very extensive cartilage, and the young probably did not move well, further supporting the idea that they were reliant on adults for support and did not forage themselves.[281]

Subsequent discoveries at the site revealed numerous other nests at the same horizon, and then also other nests with eggs across multiple horizons.[309] The obvious conclusion was that this was a major nesting site for a large number of *Maiasaura*, and also that they showed site fidelity, with animals returning year after year. Some of these nests also showed numerous eggshells and had anything from 2–24 eggs preserved within them. Others contained the remains of more juveniles, including animals as small as 0.5 m in length, likely close to hatchling size, again supporting the idea of extended care for the larger nest-bound animals.[309]

Notably, the nests were typically around 7 m apart, the length of an adult *Maiasaura*, suggesting a relatively tightly packed colony where animals stayed close to one another.[309] Such an arrangement is seen in many birds today and assists in providing defense against predators and so reduces the risk to individuals. The remains of predators that would likely have taken dinosaur eggs and young, including varanid lizards and small carnivorous theropods, have been found at the site, and hadrosaurs lack obvious defenses such as horns or armor, so this gregariousness might well have helped them protect their offspring.[308]

The remains of some eggs are sufficient to restore how they were laid down prior to hatching. These are approximately 20 cm long and were partially buried in sediment with the long axis facing upward,[309] and their anchoring in sediments means that they would not have been rotated as are bird eggs.[281] Although no remains of vegetation have been recovered from nests, adult hadrosaurs were unable to brood directly on eggs like some feathered theropods, so the eggs were presumably covered with plant material to protect them and provide heat and humidity, as with modern crocodilian nests.[309]

Overall, this clearly provides support for both pre- and post-hatching parental care, some of which was likely complex and extended for a considerable period of time, with juveniles at least doubling in length while in the nest. How much it went beyond this is hard to say, of course, although there are mixed herds of juvenile and adult hadrosaurs (e.g., [310]), so at least some might have both provided care to the minimum extent seen in *Maiasaura* and then extended it for months or even years afterward. As ever, the plasticity of behaviors means it might be excessive to extrapolate such actions to all hadrosaurs, but it certainly raises the strong possibility that such care was common in the clade.

CHAPTER 7

Combat

Perhaps almost all tetrapods will experience some form of contest or combat during their lives, and while it may be a rare occurrence, it may also be a critical one. Whether or not they have any particular adaptations for attack and defense, any such interactions, even between animals that seem ill-equipped to seriously injure one another, can result in death. As such, all dinosaurs would have engaged in some form of antagonistic interaction and physical conflict with other animals at some point, though for many this would be rare and for others the norm.

Research in living taxa tends to focus on how contests arise and the ecological drivers behind them (see [311] and papers therein) which, while important, are hard or impossible to measure in dinosaurs. Thus, the focus here is on the reality of such combats rather than their exact origins in terms of resource defense and acquisition (see [312]). It is perhaps a tautology that members of the same species have the most similar ecological demands, and thus are likely to compete with one another more than with other species for resources, be that a specialist foodstuff, a nesting site, or a territory, etc., but it is one that is worth remembering.[313] In particular, intraspecific combat, especially for the opportunity to mate, may be the difference between an animal having offspring or not. Interspecific combat, meanwhile, may involve some of the same motivations, since some species will compete with others for resources like water, and can be critical for survival. More importantly, interspecific combat can be critical to daily survival for those on the wrong end of a predator's hunger. Predation itself is covered in the following chapter since it is a component of feeding, though defense against such attempts is included below.

As noted in previous chapters, features such as weapons and armor may be multifunctional and so although selection for sociosexual signals (see chapter 5, Signaling) and intraspecific combat might have been the primary driver of features such as the brow horns of *Triceratops*, this does not mean that they could not function, or were not used, to fend off predators. Indeed, one *Triceratops* horn shows healed damage consistent with this having been caused by the teeth of a *Tyrannosaurus*,[38] suggesting at least one incident of interspecific combat between the two.

The actual mechanics of combat can involve such activities as wrestling and head-butting, as well as more obvious maneuvers such as biting or attempting to strike an opponent with feet, claws, horns, or other weapons[313] (figure 7.1). Ritualization is often an important component of combat for animals that are equipped with potentially lethal weapons, and so stereotyped behaviors can evolve where animals can assess their opponent before engaging in a full-bodied contest that could result in a serious or fatal injury.[313] This has been suggested for large theropod dinosaurs,[121] which were clearly well equipped to kill large animals. No animal wants to engage in a fight it cannot win and where there would be a serious risk of injury or even death, and so assessing the strength of an opponent and going through a series of escalations is important to give the opportunity to back down from an unwinnable situation without investing serious effort in it. This feeds back into the intersection of signals and weapons, and the importance of honest signals of animal quality.[312] Even so, there has been a recent push to separate out the two sides of this issue, since, although weapons can function as signals, they are not necessarily synonymous with being ornaments.[314]

Aggression between individuals may vary based on differences in sex and age as well as factors such as social status[315] or the time of year,[316] which will influence resource availability and events such as a breeding season. Moreover, selection can act differently on the two sexes (or even different-quality animals of the same sex), and while perhaps it seems unlikely that dinosaurs produced such diverse morphotypes as some beetles,[317] the presence of stabbing horns in female bovids for anti-predator defense vs. wrestling horns in males[318]

FIGURE 7.1 Combat between species can feature obvious weapons, such as the horns of these ibex (*Capra nubiana*), or simply whatever is available, such as the teeth and hooves of these zebra (*Equus quagga*). In either case, such combat can leave unmistakable traces, such as the facial scars on this lion (*Panthera leo*). Ibex photograph by the author; zebras courtesy of René Lauer; lion photograph courtesy of Bruce Lauer.

might be more likely (though, as noted, current evidence for dimorphism in dinosaurs is limited—see chapter 5, Signaling).

In modern ungulates, at least, there can be intersections between armaments and ecology. Horns are associated with defense of habitats or resources by males in woodland areas, showing a correlation between the morphology of male horns and differing types of habitat, social groupings, and even feeding strategies[108] (though see also [319]). Similarly, it has also been shown that in extant bovids the strength of sexual selection can be linked to the sizes of the weapons exhibited by males,[320] so there are potential correlates of major factors in biology that can be linked to the expression and use of weapons in ways that can be applied to dinosaurs.

Animals have evolved a quite extraordinary array of weapons and armaments with which they can attack other animals,[312] and this is mirrored in the dinosaurs, which display everything from the obvious teeth and claws of theropods to the "tusks" of the heterodontosaurs, horns of neoceratopsians, and the thumb spikes of *Iguanodon* and kin[321] (figure 7.2). Weapons (or at least

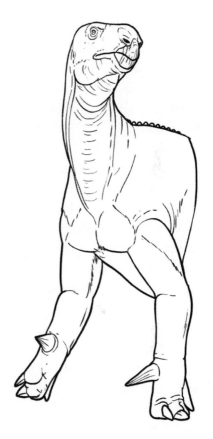

FIGURE 7.2 Life reconstruction of *Iguanodon* with the famous thumb spike on the hand. This could be used for intraspecific combat, but this has not been seriously investigated to date. Artwork by Gabriel Ugueto.

obvious ones) are rather rarer in the sauropodomorphs, though tail clubs appear in some African and Asian species,[322] the whip-tails of some of the Diplodocoidea may have functioned defensively,[265] and it has been suggested that the large claws on the hands of many might have helped serve as weapons[6] (though this might be a co-opted function, as they likely had a role in excavating nests—see chapter 6, Reproduction). Perhaps the most dramatic are the various weapons seen in the thyreophorans, with the spiked tails of stegosaurs, the tail clubs of the ankylosaurs, and most recently discovered, an axe-like set of sharp plates on the tail of the South American *Stegouros*.[323]

Although some of these features intuitively look like weapons (and all certainly could be used as such), whether or not they were functionally adapted for this or were in common use as such is a more important qualification. As a starting point, many of these dinosaurs can be reasonably inferred to have evolved these structures at least in part as weapons.

These most obvious dinosaurian weapons can also provide inferences about how these structures might have been used. For example, ceratopsians with horn shapes similar to those of some extant bovids might have similarly locked horns, or alternately engaged in flank-butting through lateral head swings,[324] and so provide important starting points for mechanical testing as well. Indeed, the various combat interactions described below are some of the best understood and most rigorously assessed behaviors for dinosaurs, because of the overlap of evidence from morphological adaptions, analogues to modern species, pathologies, and mechanical tests all pointing to similar conclusions for many species. Furthermore, studies have shown the connection between mechanical performance under loading and how weapons are used in life (in beetles, at least [314]), and this may be a source of additional data for inferring dinosaurian combat in the future.

Stabbing Weapons

Various ornithischians bore sharp weapons that would have functioned to cut and stab. The tusk-like teeth of the small heterodontosaurids are close in form to those of a number of extant deer and even some pigs, and these are used as weapons by these mammals, so this is a reasonable hypothesis for the dinosaurs too.[321] Similarly, while the horns and bosses of various bovids are not exact matches for the features of various neoceratopsians,[324] they are similar enough to infer that these cranial expansions could have been used in combat to stab at opponents. (The tail spikes of stegosaurs are clearly suited to penetrate opponents, but are considered in the following section because of their similarity in use to the ankylosaur tail clubs.)

Numerous neoceratopsians possess long and pointed horns that could have functioned as weapons, though perhaps inevitably the majority of the focus has been on the large Late Cretaceous *Triceratops*. In 2004, Andrew Farke[325] produced a hypothesis for the interactions of horns in *Triceratops*, using models to see how they might intersect in intraspecific combat (figure 7.3). This was subsequently tested by Farke and colleagues[36] by examining the pathologies of numerous *Triceratops* skulls. These crania showed numerous lesions on the same areas predicted by the manipulated models and absent in other areas—highly suggestive that these animals did engage in head-to-head wrestling contests by interlocking their brow horns. The same method was also applied to *Centrosaurus*, a centrosaurine ceratopsian that lacks the large brow horns of *Triceratops* and instead has a single long nose horn. These animals would not have been able to interlock a single horn effectively and although some show pathologies on the cranium that were likely the result of intraspecific combat they were less numerous and less regionalised on the skull, suggesting grater variance in the interactions between rivals.[36]

As noted by Farke,[321] while broken horns seen in *Triceratops* could be the result of intraspecific combat, it is impossible to demonstrate that such injuries did not occur from accidents or other behaviors. In the case of the abovementioned bitten *Triceratops* horn, the traces on the horn are consistent with bite traces attributed to the teeth of a large tyrannosaur, and as *Tyrannosaurus* is the only large carnivore in the formation, it is the only possible candidate to have left the scores in the horn. That of course implies a near head-to-head encounter between the two species and gives an idea of how the animals were positioned when the damage was made.

The claws and teeth of theropods would also provide an extensive set of stabbing and cutting weapons. Even herbivorous taxa, such as many oviraptorosaurs and therizinosaurs, often bore large and strong claws, which, while potentially optimized for digging and foraging,[326] could surely also have functioned in intra- or interspecific combat. Teeth of large carnivores remain the best candidates for this, given the presence of numerous pathologies in various species. Although known in many large-bodied theropod lineages,[37, 121] pathologies on

FIGURE 7.3 Injuries on the face of a *Triceratops* most likely caused by the horns of another during intraspecific combat. The close-up shows the elongate and worm-like scar of healed tissue above the arrows. Specimen is from the US Bureau of Land Management. Photographs courtesy of Eric Metz and the Museum of the Rockies, Montana State University.

the faces of animals that can be attributed to intraspecific biting are especially common in large tyrannosaurs. These typically take the form of subparallel score marks across the maxilla and dentary, indicating that numerous teeth raked across the face. Some of the inflicted injuries, however, are substantially more severe: in one notable case, the tyrannosaurid had part of the occiput region of the skull bitten through and removed.[327] The evidence of healed bone on these various specimens is important, as it indicates that these wounds were not inflicted peri- or postmortem, and that the animals survived such encounters—and presumably in some case, many fights. These injuries would likely have been sustained through intraspecific combat simply because there are few other credible candidates for animals that could inflict the kinds of wounds seen in these taxa. Attempted predation would likely be focused on animals much smaller than the predator, and/or on an animal that was attempting to flee (see chapter 8, Feeding). Theropod pathologies, however, are concentrated around the face, and in some cases apparently came from comparably sized animals.

Caleb Brown and colleagues[37] carried out a thorough survey of more than 200 individual tyrannosaurids (though many were represented by single elements) and found that bites were especially common on the maxilla and dentary. Estimating the size of the bite makers suggests that they were typically of a similar size to the bitten animal (figure 7.4). Importantly, the youngest specimens show no bite traces, and these increase in frequency through ontogeny, suggesting that the behavior of these animals was changing over time. Above the threshold of approximately 50 percent of adult size, lesions occur, and they continue to be acquired throughout life at approximately the same rate. Tyrannosaurids were engaging in combat at roughly the onset of sexual maturity, and they fought face-to-face. This may have been directed by courtship, territorial disputes, or general social dominance, or a combination of these.[37] Although Brown et al. did not survey the earlier-branching tyrannosauroid species, I cannot recall seeing bite marks on any of the specimens of these I have examined, and this would potentially suggest that intraspecific combat was more rare, or less focused on bites to the head, in these animals.

FIGURE 7.4 Healed pathologies on the face of the tyrannosaurid *Gorgosaurus*. The parallel lines come from multiple teeth moving across the surface of the bone and can only have been inflicted by another large theropod. Photograph courtesy of Caleb Brown and the Royal Tyrrell Museum, Drumheller, Alberta.

They do mostly have smaller skulls, with smaller teeth, and longer arms with larger claws than the later tyrannosaurs, and as the proceratosaurid groups at least had large cranial crests, they may have favored display over aggression.

Impact Weapons

An alternative to stabbing an opponent is to deliver some kind of crushing impact. Perhaps the best demonstration of this combat style are the various dome-headed pachycephalosaurs that have long been thought to have engaged in intraspecific combat (see [328]), given their thickened skulls that immediately seem well suited to ramming an opponent. Comparisons have been made between the pachycephalosaurs and modern sheep (specifically *Ovis*); despite some obvious differences, they do show key similarities[329] and might have fought in similar ways. In particular, pachycephalosaurs are suggested to have engaged in head-to-head fights, and also butted one another on the flanks or other body parts, or potentially both (figure 7.5). In either case, pachycephalosaurs show various adaptations to both the

FIGURE 7.5 Head-to-head combat between two pachycephalosaurid dinosaurs, here the relatively flatheaded *Prenocephale*. Exactly how these animals fought has been the subject of considerable discussion and remains uncertain, though some form of head-on-head impact is likely. Artwork by Gabriel Ugueto.

skull and post-cranium for delivering impacts with the dome, which strongly suggests that this was a significant part of their lifestyle.[330]

Cranial pathologies are now known for *Pachycephalosaurus* itself, and these are strongly indicative of direct head-to-head impacts, supporting this as part of the repertoire of combat in pachycephalosaurs.[331] In addition to the paleopathological data, Finite Element Analysis (FEA) of the crania of pachycephalosaurs shows that the stress absorption of the bones of their skulls would be sufficient for them to "safely" butt heads even under the highest impact regimes,[329] although at least some large bovids appear to suffer potentially serious brain injuries from such actions.[332] Notably, this is true both for animals with high domed heads such as *Pachycephalosaurus* and also for the more flat-skulled animals like *Homalocephale*. Furthermore, additional support for combat comes from other anatomical features seen in these animals. It was noted by David Fastovsky and David Weishampel[333] that the vertebral articulations would provide greater rigidity to the spine, which would assist in resisting impacts; Hans Sues[334] pointed to the orientation of the foramen magnum relative to the occipital condyle, aligning the head for impacts; and David Evans[70] suggested that the dense connective tissue that was likely present in the endocranial cavity of pachycephalosaurs would provide a cushion for the brain during impacts that would come through head-butting. Although some other features are less clear (e.g., butting bovids and caprids have hollow horns and extensive sinus cavities, whereas pachycephalosaurs have solid skulls[329]), this nonetheless collectively suggests that these animals were engaging in combat with their heads. In short, the mechanical and comparative data suggests these animals were capable of hitting things very hard with their heads, and the pathological data suggests they were hitting each other, at least on occasion head-to-head.

The tail clubs and spikes of various thyreophorans and sauropods are harder to compare to extant animals, although some porcupines will swipe their spiny tails at threats (figure 7.6). Perhaps the best analogy to these would be the glyptodonts of South America (giant armadillos that lived in the Pleistocene), some of which had clubs and spikes on their tails,[335] though of course these animals are also extinct, and thus it becomes recursive to imply that dinosaur tails were weapons based on these. However, the adaptations to stiffen the distal tail and large muscle attachments suitable for laterally directed swipes[336] make it clear that these were being swung with force, and it is reasonable to assume this was directed at an opponent (see Case Study).

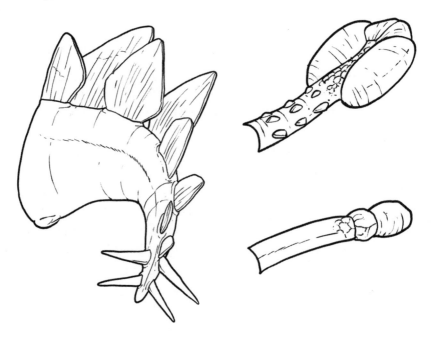

FIGURE 7.6 The tail weapons of several herbivorous dinosaurs. The set of spikes at the end of the tail of *Stegosaurus*, the end of the club tail of *Ankylosaurus* and the tail club of the sauropod *Omeisaurus*. Artwork by Gabriel Ugueto.

The armed and armored thyreophorans also provide pathology data to support the use of their weapons in combat. In *Stegosaurus*, a pathology rate as high as 10 percent was found for the spikes at the tip of the tail,[337] which strongly supports the idea that these were often impacting on other animals with a high degree of force. Although these are relatively long and thin elements that are located at the end of the tail, and thus are perhaps more susceptible to breakage than spikes or plates farther up the body, they clearly were hitting something often enough to become damaged. Although combat could be inter- or intraspecific for *Stegosaurus*, it is notable that several specimens of the large contemporaneous theropod *Allosaurus* show injuries that might have come from spike-like weapons such as those wielded by *Stegosaurus*.[337] While clearly there are other possible origins for these injuries in the theropods, it is a strong possibility that these tails were used to fend off threats, although an analysis by Victoria Arbour[336] suggested that *Stegosaurus* tails lacked the power for their spikes to puncture bone.

While pathologies both on tail clubs and on tails and pelves in general are known for various ankylosaurids, they provide limited evidence of the use of the clubs.[338] Until recently, the tail clubs of the ankylosaurs have been regarded as primarily defensive weapons used to fend off predators, and although analyses of tail club function have since led to the suggestion that these were perhaps primarily used for intraspecific combat (see Case Study), they could still have functioned effectively to attack predators. In an extensive study in 2009, Victoria Arbour estimated the impact forces from tail club strikes for ankylosaurs.[336] Some were shown to be capable of delivering bone-breaking strikes to other animals; although younger animals and smaller taxa would not have had this degree of power, even those could still deliver powerful and damaging blows.

To date, the focus on combat in thyreophorans has understandably been on the tail, given the spikes of stegosaurs, the clubs of ankylosaurs, and the axe of *Stegouros*. However, while *Stegosaurus* is famous for its spiked tail and otherwise has a row of plates along its back, most stegosaurs have spikes up much of the body, rather than plates, and also have a pair of large parascapular spines that extend out from the shoulders. The nodosaurid ankylosaurs lack tail clubs but also often have large spikes around the shoulders, and a number of early thyreophorans such as the polacanthines have very spiky armor. These various bony protrusions may have had a combat role in some kinds of ramming aggression between conspecifics, or even potentially against other species.[339]

Turning to the saurischians, a number of theropods have been suggested to have fought one another by butting with their heads, with the various hornlets and the bosses providing a weapon for doing so.[340] This hypothesis has ranged across numerous theropod clades, but has focused especially on large-bodied and thick-headed animals such as the derived tyrannosaurs and abelisaurs. The horns or hornlets of various species are at least somewhat similar to the horns of some caprids and are a viable candidate for another mode of intraspecific combat. Gerardo Mazzetta and colleagues[341] examined the possibility of head-butting in the abelisaurs, and their work using Finite Element Analysis (FEA) suggested that the South American *Carnotaurus* could not have endured severe, rapid frontal blows during agonistic encounters. This does not rule out head-butting in these animals as part of their intraspecific combat repertoire, but it does suggest that if head-butting did happen it was, at most, an occasional event (in contrast with pachycephalosaurs), and that the horns were predominantly ornamental rather than adapted for impacts.

One final example to consider in this area is the suggestion that some of the apatosaurine sauropods engaged in a form of neck-to-neck combat, as seen in animals including giraffes and elephant seals.[342] Noting the bulk of these animals, and the unusual knob-like expansions on the underside of their cervical ribs, suggests that the latter could have been used in combat. This hypothesis has yet to be fleshed out in detail, but remains a plausible explanation, supported by the morphology of robust cervical ribs in these animals, which would have increased the strength of the neck when moving ventrally.[342] Certainly this can be tested by examining the mechanical strength of these bosses, looking for pathologies on the cervical ribs, and investigating the ontogeny of these expansions. Finally, in the sauropods, a tail club referred to the Middle Jurassic *Mamenchisaurus* was shown through FEA analysis to have had limited function as a weapon.[343] These tails lack the stiff handle of ankylosaur tails; the club is a simpler structure of a few fused caudal vertebrae, not the complex one of ankylosaurs, and would not have functioned as effectively. It was shown to be more effective swung laterally than dorsoventrally.

Summary

Although there is clearly some strong evidence for both inter- and intraspecific combat in the Dinosauria, across the clade as a whole the data is rather limited. To date there has been a strong focus on a small number of taxa which bear likely-looking weapons, with other groups almost entirely ignored. The thumb spikes of the *Iguanodon* and its relatives are often regarded as weapons, and have been illustrated being used as such in innumerable books, but there has been no investigation into how they might have functioned, and no reports of the kinds of pathologies they might leave on each other or on contemporaneous theropods. Similarly, the prosauropods lacked the speed to evade theropods, were too big to easily hide, but lacked the huge bulk of the sauropods that might give them immunity to predation. Put together, that would leave them very short of any kind of protection, and so suggests that their relatively large and curved manual claws might have been their best defense against predators—but that possibility has never been looked at seriously. Despite their diversity and huge sizes, the sauropods themselves have similarly undergone little scrutiny into how they might have fought one another or potential theropods, and this is another area ripe for further consideration.

Even for the neoceratopsians, with their plethora of facial horns, studies have focused primarily on *Triceratops* and other taxa such as *Centrosaurus* (single large nose horn) and *Pachyrhinosaurus* (giant nasal boss), despite their very different armaments, while very large numbers of other species have had little attention. Various early ceratopsians lacked horns or bosses entirely, but had very sharp beaks that could have functioned as a weapon; but again, these have not been assessed as such. Similarly, in the theropods work has been concentrated on the larger species, although the various smaller forms (and especially feathered species) must also have fought one another on occasion and were well equipped to do so, with sharp teeth and claws on both hands and feet.

Combat must have been important for almost all species, even if apparently ill-equipped, but inevitably evidence will be limited. For example, for all the huge raptorial claws on the feet of the dromaeo-saurs, these puncturing weapons, coupled with relatively small teeth, might not leave such obvious scars and marks on each other's skele-tons as tyrannosaur teeth will; and hadrosaurs body-slamming one another might produce some broken ribs and tails, but injuries sus-tained in this manner of combat (if it even happened) would be hard to separate out from other damage. Therefore, while comparisons to extant taxa are useful starting points for hypotheses, and pathologies can be highly informative, the development of more biomechanical analyses of dinosaur weapons and interactions are likely to be the most important area of study going forward to determine how these animals fought one another.

CASE STUDY: **Combat between ankylosaurs**

Although many animal weapons are quite apparent from their overall morphology, few in dinosaurs are perhaps as obvious as the tail clubs seen in various ankylosaurids. These consist of a series of relatively normal vertebrae at the base of the tail, followed by a set of heavily overlapping bones with V-shaped neural spines that form a rigid "'handle" (that is also supported by extensive clusters of ossified tendons), and terminate in a cluster of osteoderms fused around the distalmost vertebrae to form a solid club.[336] Larger tail clubs could be over 30 cm across and in life would have weighed tens of kilos. Coupled with extensive muscle mass from the broad pelvis, these clubs would be swung with considerable force, estimated at 7,000–

14,000 N, to deliver an enormous impact: enough to break the bones of a potential assailant.[336] The clubs are well suited to this activity, and pathologies in ankylosaurid tails are consistent with their absorbing major impacts.[338]

The classic interpretation of these weapons is that they were used in conjunction with the extensively armored body of ankylosaurids to defend themselves against predators, and in particular the large tyrannosaurs of the Late Cretaceous with their very strong bites (figure 7.7). However, some obvious questions remain about this interpretation. If the tail club was so important as a defense, why is it missing in the nodosaurids (the sister taxon to the ankylosaurids)? Presumably the similar levels of armor, and therefore similar protection, present in the nodosaurids was sufficient to keep tyrannosaurs at bay, so then the tail appears to be an unnecessary encumbrance rather than

FIGURE 7.7 Part of the armor of the ankylosaurian *Zuul* showing the numerous osteoderms that would protect the animal from injuries, be they from attacks by predators or conspecifics. Specimen at the Royal Ontario Museum, Toronto, Canada. Photograph by the author.

a benefit. If so, what accounts for its presence? Similarly, juvenile dinosaurs are typically the most vulnerable to predation,[88] and yet young ankylosaurs lack clubs (and have reduced or no armor), which would mean they had no weapons to defend themselves.[336] Finally, many animals employ their tails for defense against competitors and predators, but relatively few are equipped with any kind of specific weapon.[335] All this points to an alternative explanation for tail-clubbing behavior—namely, intraspecific combat. Ankylosaurid tail clubs evolved to hit other ankylosaurids, not tyrannosaurs.

This then may also provide an alternate explanation for the armor itself. If ankylosaurs were swinging their tail clubs at each other, extensive armor would have helped protect them from the blows of their fellows. The tail clubs would swing laterally, and the armor of ankylosaurids matches this general orientation, with the largest plates and osteoderms on the flanks, sticking out laterally (hardly the orientation to deter a large theropod striking down from above). Evidence of pathologies from heavy impacts are now known in ankylosaurs, further supporting intraspecific combat as a potential driver of both armor and clubs,[344] and this seems likely to be the most important factor in their origin and expansion. Doubtless the armor did help defend against predators—and this was likely a common driver in its origin[335]—and the tail club would have been used to fend of attackers of all kinds (indeed, a number of tyrannosaurs from ankylosaurid-preserving deposits show lower-leg injuries that are rare in other large theropods). However, the origin and driver of the club specifically might well have been intra-, not interspecific, combat.

CHAPTER 8

Feeding

As with reproduction, the idea that animals need nutrition and energy is such a fundamental aspect of biology that there is perhaps little that can be said in the way of an introduction to this subject. All dinosaurs would need to find, obtain, and process food, though there are massive differences between browsing sauropods, macropredatory theropods, and everything in between. This is (alongside combat) one of the most studied areas of dinosaur ecology and behavior, since numerous lines of data are available (jaw and tooth shape, gut contents, tooth microwear, isotopic signatures[345]) that can be aligned to produce strong evidence for some basic actions. Even so, the diet of numerous lineages is unknown, or inferred from only basic comparative anatomy of, e.g., tooth shape in herbivores and carnivores; indeed, studying the diet of even living animals is actually very difficult. Working out what they eat, when, and why, and all the huge variation in foraging and consumption (accidental or deliberate, occasional or normal) can make for some major surprises. We can reasonably infer that a given dinosaur was a herbivore, and its size and tooth shape might point it out as a bulk feeder, but when and where it foraged, and what plants were preferred, and how this might vary over seasons or years and in different habitats, will be effectively unknown. This chapter, however, is not a list of which species ate what, but is about the hows and whys of foraging and feeding.

A major aspect of feeding ecology lies in the energetic trade-offs and risks of different behaviors. Animals will inevitably prefer high-quality food patches over low, and low-risk areas over high (see [346]), but such areas will vary constantly with depletion of resources, competition for food, travel time between sites, predator numbers and their habitat preferences, and the like. As ever, perhaps the best thing we can do is to look to such basic patterns of foraging trade-offs to make general inferences about how dinosaurs behaved (often linked to fundamental size), bolstered by hard evidence in specific cases where the fossil record allows.

For example, larger species (and individuals) need more food than smaller ones. However, physiological efficiency goes up with large size,[347] and so absolute requirements will increase, but the proportional requirements will be lower. Larger animals also travel more efficiently[347] and so can move between resource patches more effectively, and also will occupy larger areas. This means that they will encounter more varied environments and conditions (and likely foodstuffs) and might be required to move to find food, so at some level, inevitably, to be more of a generalist, where smaller animals in smaller areas can potentially be specialists (figure 8.1). Mouth width correlates with dietary preferences (in ungulates, at least[348]), and shows the general split of herbivores into bulk feeders (and digesters) who consume any and all foods in large quantities to digest, and those selective feeders targeting a smaller volume of high-quality foods. Herbivores cannot forage and be vigilant simultaneously (also a major driver of group living, see chapter 4), and the very act of foraging and feeding makes animals more vulnerable.[349]

Large carnivores have a potentially wider range of prey species than do smaller ones (since large animals can tackle both large and small prey, but small predators are limited to smaller prey), although they do not always take advantage of this.[350] Predators tend to be either pursuit predators—which actively chase prey, either at high speeds or across significant distances—or ambush predators, which rely on getting very close to prey before striking,[351] but each includes elements of the other: ambushers still need to chase prey, and pursuit predators benefit from reducing the distance before striking. Group living by predators can have important influence over their behaviors, with potentially higher success rates in larger groups and the ability to tackle larger prey (see [352] on spotted hyena, *Crocuta crocuta*), though the trade-off for cooperation is that the captured prey then needs to be shared, reducing intake. Carnivores are generally less active than herbivores, taking high-quality food that is easy to digest (meat),

FIGURE 8.1 As large animals with a broad distribution and large territories, African lions (*Panthera leo*) can be found in a diverse set of habitats, from open savannahs to dense bush and, in this case, desert. This contrasts with animals like dik-dik (*Madoqua saltiana*) with a much narrower diet, range, and habitat occupation. Lion photograph courtesy of Robert Knell, dik-dik photograph by the author.

and it means that active hunting can be a relatively small percentage of their activities (figure 8.2).

Numerous similar species of dinosaurs are often present in the same ecosystems (or at least fossil localities),[4] and this is hardly novel—this is seen in both extant and ancient ecosystems, even when

FIGURE 8.2 Nature is not constantly red in tooth and claw. Here two cheetah (*Acinonyx jubatus*) are not actively hunting and so can be largely ignored by the warthogs (*Phacochoerus aethiopicus*) even though they are potential prey. Photograph courtesy of René Lauer.

it comes to large predators.[353] There must have typically been niche separation in order to maintain stable populations of these taxa, as Gause's axiom states that species with identical niches cannot coexist, so one must be outcompeted, move, or adapt.[354] Similarly, dietary overlap is common, for example between animals as different as lion (*Panthera leo*), cheetah (*Acinonyx jubatus*), and wild dog (*Lycaon pictus*) all preying on a number of the same species of antelope.[355] Despite this, they do occupy separate niches, so the differences in what else they take, or which parts of the population they target (e.g., juveniles or adults), are the important aspects of niche separation, rather than the methods by which they hunt or feed. Studying dinosaur ecology, therefore, especially as it related to feeding, should be focused on how animals might have separated their niches rather than simply restating that niche separation must exist, or that differences in feeding or hunting style automatically indicate niche separation. Guilds of similar feeders would likely have had at least some overlap, and again large animals will tend to be generalists rather than specialists; this should be the starting point of investigations, not the endpoint.

Happily, this is increasingly being taken up by paleontologists, with, for example, isotope data used to demonstrate different plant feeding preferences in sympatric hadrosaurs and ceratopsians, and so pointing to the mechanism of niche separation in these taxa[22] despite their overlap in potential feeding ranges.[356]

Many, if not all of these patterns would also be true for dinosaurs and at least some are detectable in the fossil record, and help unveil how dinosaurs foraged and fed and what drove their behaviors. Finding and processing food would have been a major component of the daily lives of dinosaurs, and we would expect it to exert a major selective force on both anatomy and behavior. At least some of the latter can be pieced together with the former.

Herbivores

Although considerable work has been done on the ecology and feeding of herbivorous dinosaurs (see [16]), much of this can be simplified given the balance that is perhaps inevitable for moderately sized through to large-bodied herbivores. Plants are generally not very nutritious compared to animal protein, and so require long digestion times to break down, which is easier for larger animals with longer guts. Therefore, large animals will tend to be bulk feeders, taking on as much food as possible however poor in quality, while smaller ones are capable of being more selective, focusing on buds and shoots, which are more nutritious and require less digestion time (see [108]). Thus body size alone will largely determine how much food and what type (small shoots, buds, etc., or whole leaves and perhaps twigs and even branches) herbivores favored. Similarly, selective feeders on highly nutritious resources will require far less time consuming and processing food than large animals, and, being small, may be able to hide more effectively and move less between resource patches, driving other key aspects of behavior such as vigilance and territory occupation (see chapter 4, Group Living). Ultimately, this means that large herbivores can survive on poor-quality food, but small ones require less food.[132]

Mouth size and shape has therefore been used as a starting point for inferring the general diets of herbivorous dinosaurs, and while some broad strokes can be easily made (e.g., a selective diet for small taxa like heterodontosaurs and many juvenile ornithischians; bulk feeding for large animals such as the giant hadrosaur *Shantungosaurus*), some specific work has also been done. For example, the jaws of ankylosaurs have been shown to widen during their evolution, suggesting a shift from a more specific and selective diet to more general consumption of plants.[357] In the hadrosaurs, the saurolophine group have narrower beaks anteriorly, while the lambeosaurines show wider beaks, pointing to a similar split in approach to diet[358]—although their dental morphology is strikingly uniform in the two clades, implying that they were processing foods in similar ways.[109] Interestingly, these findings can be combined with data on limb lengths: hadrosaurs from coastal regions show shorter forelimbs than animals from more inland settings, suggesting that the coastal dwellers were more bipedal and therefore perhaps foraged at higher levels.[358] A related major line of strong data on food preferences comes from tooth wear for large herbivorous ornithischians in the Cretaceous, indicating that ceratopsians favored fibrous foods, hadrosaurs were capable of taking a wide range of plants (figure 8.3), and the armored dinosaurs split, with nodosaurids preferring more fibrous food than did the mixed-feeding ankylosaurids.[359]

FIGURE 8.3 The dental battery of the north American hadrosaur *Parasaurolophus* shown in medial view. The enormous rows of teeth move up the jaw as they grow and are worn away rapidly at the top as they grind against the teeth in the upper jaws. Photograph courtesy of Ali Nabavizadeh.

Stomach contents and coprolites are rare for herbivores (given the lower preservation potential of digested plants compared with bones consumed by carnivores), but some are known, including gut contents of a nodosaur showing preferential consumption of ferns[360] and hadrosaur coprolites showing they (at least on occasion) ingested wood.[361] In the latter case, there is also evidence of the consumption of crustaceans that likely sheltered within rotten wood resting in water, though whether this was accidental or targeted consumption is not known.[361]

In the case of the large sauropodomorphs, adaptations for simple cropping of vegetation coupled with the lack of oral processing (figure 8.4), long digestion times, and potentially high energetic returns meant that they could have easily consumed sufficient vegetation to support themselves.[362] Differential feeding is clearly demonstrated in at least some sympatric animals, as exemplified by the more powerful bites of *Camarasaurus* vs. *Diplodocus* in the Late Jurassic Morrison Formation, pointing to specializations in different plants[363] and therefore niche partitioning. Collectively, though, the foraging behavior of sauropods is complex and controversial, especially with respect to the function of the long neck in acquiring food (see Case Study). For some large sauropods[137] and ornithischians,[138] we have strong evidence for

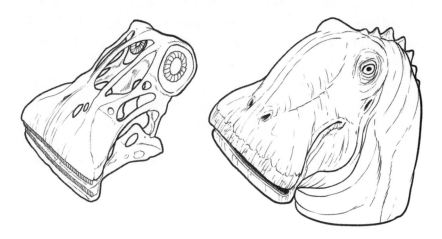

FIGURE 8.4 Skull and life restoration of the small Early Cretaceous diplodocoid sauropod *Nigersaurus* from Niger. The numerous teeth in a broad jaw and relatively short neck point to an animal that was a low browser or grazer. Artwork by Gabriel Ugueto.

migratory behaviors, and this could well be linked to diet, given the inevitable need to follow changing weather patterns over an annual cycle to find optimum foraging conditions.

Ontogeny would inevitably be at play, with young animals unable to reach the heights of adults, or to have the bite force or digestive time to effectively process more fibrous and tougher plants (see [364] on sauropods and [209] on hadrosaurs). The diets of juvenile animals have been studied far less than those of adults (perhaps inevitably, given the general rarity of juveniles). One interesting exception is the small Middle Jurassic theropod *Limusaurus*, whose adults have beaks and were herbivorous, while the juveniles were toothed and were probably omnivorous,[47] or at least must have had a diet distinct from that of the adults.

One notable feature of herbivorous dinosaurs is the presence of a beak. Although this was universally present in ornithischians, beaks appeared independently multiple times in theropod lineages which became herbivorous, and this was clearly a common theme in processing plant material. Even more notable convergences occurred between various lineages: the skulls of Triassic prosauropods, Jurassic stegosaurs, and Cretaceous therizinosaurs show a gross similarity of form, although mechanically they performed differently, and the three clades likely ate different things.[365] For example, the downturned jaws seen in therizinosaurs are linked to their high bite efficiency,[366] which implies a preference for tougher plants (figure 8.5). Similarly, the jaws of various oviraptorosaurs are suggestive of high bite forces (both absolute and relative to other herbivorous theropods), as determined from analyses of their jaw muscles in both early toothed and derived beaked taxa.[367] Coupled with a smaller gape than is seen in predatory theropods, this suggests, again, a preference for tough plants.[367] It has been suggested that some early ornithomimosaurs were filter feeders (figure 8.6), but the energetics of such a diet have been shown not to be viable, and based on the available data they were most likely primarily herbivorous.[368]

One other factor that occurred with many of the beaked theropods was a need for gastroliths (stomach stones), ingested to help grind up food in the absence of teeth, and these are notably absent in the

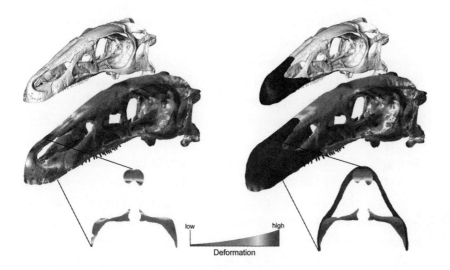

FIGURE 8.5 Finite Element Analysis of the stresses on the skull of the therizinosaur *Erlikosaurus* showing the stresses passing through the skull during feeding; this pattern is convergently similar to those in both a prosauropod and a stegosaur. Graphic courtesy of Stephan Lautenschlager.

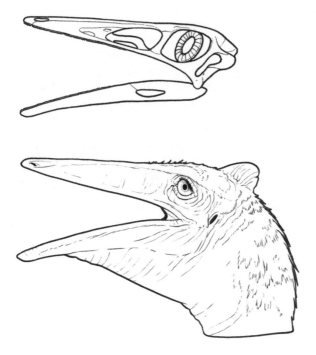

FIGURE 8.6 Skull and life restoration of the Spanish ornithomimosaur *Pelecanimimus*. The diet of this animal remains controversial and the odd combination of thin teeth and pelican-like (if only in appearance and not function) throat pouch adds to the uncertainty. Artwork by Gabriel Ugueto.

ornithischians and sauropodomorphs. Gastroliths were first reported in oviraptorosaurs[369] and ornithomimosaurs,[370] and are now known in various other herbivorous theropods as well (including herbivorous ceratosaurs). These are analogues to those of some modern birds, though theropods lacked a specialized crop, and gastroliths would have been kept in the stomach. Exactly how these were selected has not been studied, though they appear to have worn down quickly in at least some cases and would have required regular topping up.[370]

As bipeds, the theropods in particular had the forelimbs free and were potentially able to use these to procure food. The claws of some therizinosaurs have been shown to be potentially effective both in digging and potentially in hooking vegetation when foraging,[326] and this has also been suggested for the giant ornithomimosaur *Deinocheirus*[371] (though this animal may have been omnivorous). Quite how the claws would have functioned, however, and what plant parts (leaves or roots) would have been targeted, requires further investigation.

Carnivores

Considerably greater attention has been paid to food acquisition and processing in theropods (see [345]), in part, perhaps, because of the popularity of these animals as carnivores, but also because the range of adaptations and possibilities for acquiring food are so much greater for them than for herbivores (since plants generally cannot run away). Carnivores also have the options of predation (actively killing and consuming animals) or scavenging (feeding on animal material that is already dead) or both, the latter being by far the most likely course of action for most carnivores. At least some of the earliest ornithischians and sauropodomorphs may have been carnivorous or omnivorous,[16] although little work has been done on these; the same principles would apply in terms of the carnivorous parts of their diets.

Although stomach contents are known for numerous theropods,[372] evidence for predation vs. scavenging is very limited. Demonstrating each behavior requires specific evidence: for predation (or at least attempted predation), strikes on bones that have subsequently healed (though some are known: see [113]) to show that a prey animal was alive when attacked; for scavenging, evidence that an animal was dead and had undergone a process such as transport prior to consumption (e.g., see [114]), and so this was likely beyond simply late-stage carcass consumption by a predator. In this regard, the term "carnivore–consumed" should be preferred to the often used "predator–prey" when discussing these interactions, since it is not known or clear in most cases if active predation was involved.[327] As with modern carnivores, most carnivorous or omnivorous theropods would have likely engaged in both, though to what extent is not clear. The only case ever seriously made for pure or predominant scavenging was for *Tyrannosaurus*, and this is not well supported at all.[373] What can and should be primarily considered is what might have been taken and how.

Theropods likely focused predominantly on animals much smaller than themselves, and in particular, juvenile dinosaurs. As in modern animals, juveniles are typically much smaller than adults, lack defenses like armor and horns, are naïve about the threats of predators, and often need to forage for longer durations in low-quality areas, all of which makes them a preferential target (see [88] for a review). This is reflected in the fossil record with the increased tendency of juvenile dinosaurs to form groups (when adults might not—see chapter 4, Group Living); the presence of healed injuries on juvenile dinosaurs that are absent on adults; and the rarity of juveniles in the fossil record (because they were consumed and destroyed), even allowing for the reduced preservational potential of incompletely ossified bones.[88]

Note that juvenile dinosaurs might still be relatively large (even a partially grown *Triceratops* or *Edmontosaurus* below sexual maturity would be over a ton), while smaller theropods would have likely taken non-dinosaurian prey, e.g., as seen in the tiny theropod *Scipionyx*, whose gut contents show that it consumed insects as well as small fish and lizards[374] (the overall trend to tackle prey much smaller than the predator is again a typical one[372]). At the other end of the spectrum, large sauropods have been considered largely outside of the range of prey for theropods. Mark Hallett and Matt Wedel[6] suggested a number of ways that theropods could have attacked even large sauropods (e.g., attacking the Achilles tendon, or repeated bites to weaken the animal through bleeding), pointing out the potential for groups of theropods to cooperate in targeting larger animals and the apparent correlation of increasing predator size with prey size in mammals. However, this need not conflict with the central preference for smaller and/or juvenile prey—after all, a half-sized titanosaur would be larger than a half-sized diplodocine, and so might select for larger theropods to tackle them or to operate in groups, still without ever favouring adults as a target.

Records of interaction between carnivores and herbivores, including sromach contents, have been used to argue for prey preference. For example, the famous specimen of a *Velociraptor* and *Protoceratops* apparently locked in a death embrace has been used to argue that the ceratopsian was the preferred prey of the dromaeosaur.[170] However, this is a singular incident and a failed predation attempt (as the *Velociraptor* died), and so might represent an abnormal situation, not a typical prey choice. Animals that were more commonly fed upon will be more likely to turn up as stomach contents, though not always, as for example feeding on pure muscle or on soft-bodied invertebrates will not likely show up in the fossil record as preserved stomach contents compared to bone.[372] In some cases, there is sufficient data to show clear patterns, such as the numerous bite marks attributed to large tyrannosaurs seen on ornithischian bones in the Late Cretaceous Dinosaur Park Formation of Canada, which are a strong match for the approximate numbers of hadrosaurs and ceratopsians in the formation, suggesting that tyrannosaurs fed on these according to their local abundance rather than showing any clear preferences for specific species.[375] Alternatively, multiple examples of gut contents for the small gliding *Microraptor* show that they ate mammals, birds, lizards, and fish and suggest a generalist, rather than specialist diet.[372]

The crania of theropods are key to their habits, housing the major senses and the jaws and teeth to catch and process potential prey, and various traits provide strong indications of what they were doing and how. Although large tyrannosaurs have the best binocular overlap and largest eyeballs of any theropod, and thus perhaps the best overall vision, theropod skulls in general show that they were binocular and would have been good at judging distances, a key part of predatory behavior.[61]

More important, though, is the ability of theropods to bite into potential prey or carcasses when feeding. Bite forces overall can be calculated or estimated effectively[376] for theropods, and large tyrannosaurs again have the most powerful bites known, with robust teeth, strong jaws, and reinforcements to the skull to deliver powerful bites that can puncture bone.[377] Bite strength increases in a number of theropod lineages over time, and within the tyrannosaurs ontogenetically.[366] This suggests that early theropods had a weak bite and favored even smaller prey than did later and larger taxa.[366] Variation is seen, however, with, for example, the abelisaur *Carnotaurus* (figure 8.7) shown to have had a weak but fast bite, combined with a skull that was capable of withstanding high-velocity impacts, suggesting the head might have been forced into the target to produce a stronger bite.[341] Within the dromaeosaurs, those with a longer snout like *Velociraptor* had a quick but light bite and likely favored smaller

FIGURE 8.7 Skull of the abelisaurid theropod *Carnotaurus*. Analysis suggests that this tall and thin skull was well adapted for fast, but light, bites. Photograph courtesy of Christophe Hendrickx.

prey,[378] which again argues against tackling subadult *Protoceratops* as a typical target.

Feeding is a separate issue from prey capture. Theropod stomach contents show that small animals were consumed whole or in relatively large parts, and feeding on larger bodies would require alternate strategies. Bite marks on bones are rare for most theropods (though more common in tyrannosaur-dominated environments), suggesting that tooth on bone contact was generally avoided by most theropods despite their capacity to bite through bone on occasion.[88] This, in turn, implies that they fed primarily on muscle tissues and did not habitually break into, or bite, bone—feeding behavior similar to that of other large carnivores that have stereotyped carcass consumption patterns targeting areas of high muscle mass first (see [379]).

Despite the higher incidence of bone bites inflicted by tyrannosaurs, perhaps even they were not dedicated bone biters. Bite marks on a hadrosaur humerus by a large tyrannosauine show that their specially flattened premaxillary teeth were used to scrape muscle tissues from the bone, and while this left numerous traces, they did not simply bite through the thin deltopectoral crest, which suggests that they were relatively sophisticated and selective in feeding.[114] This also fits with studies of the neck musculature of large theropods that shows that they were well suited for a feeding style of retracting the head in the manner of modern birds of prey,[380] and these muscles would have been important for scrape feeding, pulling meat from bones, and pulling apart a body (figure 8.8). In this latter case inertia might be a problem, but the carcass could be held down by the feet. The slight asymmetry seen in the grooves of the second pedal ungual of many predatory theropods would make this bone stronger (increasing the space between the grooves and so reducing the chance of its breaking), and suggests that this bone took more force than other claws, which would be the case if the feet were used to pin down a body while the head was employed in biting.

Theropods were not simply heads, however, and the rest of their bodies would have contributed to their predatory behaviors. For example, multiple, and often serious, pathologies to the arms and hands of a specimen of the Early Jurassic *Dilophosaurus* point to heavy use,[39]

FIGURE 8.8 A griffon vulture (*Gyps fulvus*) feeding in a manner that was likely typical for many theropods, with the foot used to pin down the food item and the head being pulled back to rip it apart. Photograph by the author.

and arm pathologies are known in a variety of theropods. Quite how the arms were used in various clades is unclear. In most of the large predatory theropods, the arms were strong but also not especially long, so they would be under the body, out of sight of the animal for manipulating things, and the head would reach any potential prey before the arms when striking at something ahead. In short, the fore-limbs were not obviously well suited to catching or manipulating prey, though their musculature and pathologies suggest they did have important functions (intraspecific combat is at least one alternate option). In contrast, long arms are seen in various predatory lineages close to birds, coupled with hands that appear well suited to grasping things:[149] sharp claws and deep ligamentous pits on the phalanges point to a strong grip, suggesting a key function in predation.

In these bird-like dinosaurs, the feathers themselves might also have played a role, with the suggestion that the long pennaceous feathers on the hands of dromaeosaurs could have been used to help control small prey.[381] This action is termed "mantling" and is based on the habits of many modern birds of prey, who encircle prey with the wings to help prevent their escape; a similar action might have been undertaken by some theropods.[381] This might also match with the shape of the pedal claws and their ability to pierce and potentially hold prey, rather than being used for slashing, as once thought.[382] The feet of the dromaeosaur *Microraptor* have been likened to those of modern predatory birds, in contrast to many other early flying non-avialians, suggesting an important role for the feet in prey capture and/or processing,[383] while observations of the extant predatory seriemas (*Cariama*) support the use of the feet in piercing and controlling prey for dromaeosaurs, at least.[384]

The legs of theropods would also have been important, and the general abilities of theropods to accelerate and turn, and their maximum speed and potential endurance, would all be factors in how prey was taken (see [154] on speed, [116] on turning, [385] on acceleration). Theropods would have employed fundamental strategies similar to those of other predators, and so would mostly either have given chase over a longer distance, relying on endurance to wear down prey (likely employed by tyrannosaurs[386]), or run quickly in an attempt to capture them using speed (likely favored by abelisaurs[385])—though in both cases approaching prey relatively undetected would have improved their chances of a successful hunt. Important data on these strategies also comes from the potential prey themselves, and their abilities to accelerate and attempt to avoid predators (see [386] on hadrosaurs), but this is an area that requires more attention in the future. Attacks would likely have predominantly come from behind as the result of some kind of chase, and this hypothesis is supported at least in part by the number of healed injuries seen to tails and hindlimbs of herbivores. The large caudofemoralis muscles would have been an obvious target, and debilitating if bitten into.[117]

The generally tiny alvarezsaurs have an unusual set of features that actually link strongly to an insectivorous diet. The small and toothless head, combined with large manual claws and a robust manus and elongate olecranon, point to animals capable of digging into hard substrates to access small and relatively soft foods such as termites or beetle larvae,[156] and recent functional analyses of the claws suggests that they were indeed well suited to digging.[387] Other anatomical details also fit this putative lifestyle, with, for example, the presence of a stiffened spine (which would assist in digging) and restricted middle metatarsals (similar to those of tyrannosaurs and others) suggesting that moving efficiently was a key part of their lifestyle—as it is for some ant and termite specialists today[107]—while their asymmetric ears could have been useful for operating at night.[62]

Another clade of unusual predatory theropods are the spinosaurs; numerous papers have been published in recent years on their putative behavior and predatory habits. See the Case Study in chapter 3, The Basis of Dinosaur Behaviors, for more details.

Ontogeny would also be in play in carnivores, on the same basic premise as in herbivores: juvenile animals that were perhaps 0.01 percent or less of adult mass (in the case of giant tyrannosaurs or allosauroids) can hardly have been targeting the same prey in the same way as adults, nor would they have been able to process carcasses that adults could. This becomes clearer with, for example, the changes in tooth shape, skull size, and leg length as the juveniles of large tyrannosaurs matured, indicating that young animals shifted niches quite significantly as they grew (probably multiple times in the case of very large taxa), and likely took in different foods in fundamentally different ways than adults.[49] Evidence for diet shifts are also seen in dromaeosaurs, with the North American *Deinonychus* showing differences in isotopic signature of tooth enamel of older and younger animals—suggesting different diets for the two, and also arguing against multigenerational group hunting.[388]

Summary

The basic diets of many dinosaurs are understood and well supported across multiple lines of evidence. Although detailed studies using biomechanics, trace fossils, stomach contents, comparative anatomy, and other lines of data are limited to a handful of species (see [23]), these are often well-founded, with numerous different lines of evidence pointing to the same general conclusion. However, even in cases such as the tyrannosaurs that have been the subjects of numerous papers on their hypothesized habits, large areas of uncertainty persist, or there remain conflicting accounts and interpretations. Of course, given the variation in ontogeny, habitats, differening prey species, or simply individual variation, there is likely no singular "right" answer for how they hunted and fed.

Even so, the coarse levels of these studies do align well, and more-over we have some excellent details for a variety of species, such as the diverse stomach contents seen in *Scipionyx*, or the evidence for specialist feeding with the premaxillary teeth in large tyrannosaurs. In the case of herbivores, detailed work from stomach contents, isotope data, and enamel scratches can point to specific diets, and comparisons between species show up clear differences that point to niche partitioning. Bridging the gap between the gross generalities of "ornithischians were herbivorous" and "adults of this species likely focused on low ferns in this environment" is the challenge going for-ward. The flexibility of diets, and perhaps the inevitable generalist foraging approaches of large dinosaurs (both herbivores and carni-vores), mean that many gaps will be difficult to fill in. The increasing amounts of data available at least help to demonstrate what animals fed on and to a degree, how, even if more specific foraging behaviors and preferences, and the ecological interactions between taxa, remain somewhat cryptic. Collectively, we have superb data on dinosaurs feeding, and this should be an area where this baseline information can be the foundations of much more detailed studies of specific behaviors moving forward.

CASE STUDY: **Feeding height in sauropods**

The discovery of the first sauropods led to the inevitable ques-tion: What was the purpose of such an extremely long neck? These are truly extraordinary, with many of them exceeding 10 m in length.[18] Giaffes with theis ability to reach high into trees might have been the most obvious comparions, but in the context of a time where dinosaurs were seen as some kind of failed evolutionary experiment, lizard-like animals with a sprawling posture and limited abilities, it is perhaps unsurprising that the idea soon became set that they were feeding low down. Specifically, sauropods were considered aquatic animals using the water to support their weight and feeding on subsurface water plants, so their ability to reach their necks down would have been useful. This idea has in some ways never entirely gone away,[389] even though it is clearly incorrect—not least because few plants grow at any depth due to the lack of light and that sauropods, and especially their necks, would have been highly buoyant because of the air sac system.

A common recent alternative has been the suggestion that sauropods engaged in a more goose-like approach to foraging, with their long necks allowing the animal to reach out and sweep across a large area,[390] on the principle that a heavy animal can then minimize its movements while feeding. Although there is some support for this idea from the arrangement of sauropod cervical vertebrae and muscles,[391, 392] it has also been noted that the apparent energy savings from this model are limited, since larger animals would be more efficient locomotors than smaller ones, and by grazing or browsing low down they would not escape competition with other species[6] (or juvenile sauropods), so this was likely not a universal selection pressure. That said, evidence of grasses has been found in titanosaur coprolites, so at least some sauropods were feeding low down at least part of the time.[393]

In the case of the famous North American Jurassic sauropod *Diplodocus*, cases have been made for both high browsing in trees (figure 8.9) and relatively low browsing or even grazing. Paul Barrett and Paul Upchurch[394] argued for a branch-stripping model of high feeding, in which branches of plants like tree ferns could be taken into the mouth and the head pulled backward to strip off the leaflets. This was supported by the wear patterns seen on the teeth and the high flexibility

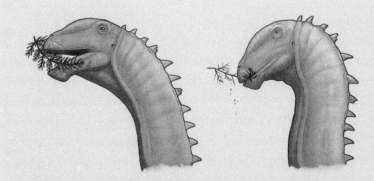

FIGURE 8.9 Hypothesized feeding technique of *Diplodocus*, based on [394]. A series of leaves and twigs are bitten and then the head and neck pull back, stripping these from the tree and leaving wear on the medial faces of the teeth. Artwork by Gabriel Ugueto.

of the head-neck joint angle, though they noted that some low browsing also likely took place, based on wear on the lower teeth, and that the short forelimbs could have helped to keep the head close to the ground (though it does seem that sauropods had a habitual high-neck posture,[146] cf. [390]).

Some later studies have supported this high-feeding model too: for example, Finite Element Analyses of the skull of *Diplodocus*[395] demonstrated that it exhibited low stresses under nipping and branch-stripping actions, suggesting that it was well optimized for such behaviors. It has also been noted that these are "hindlimb dominant" animals, with relatively large and strong hindlimbs and the weight concentrated to the rear, and would be well suited for rearing up, pushing the head high off the ground.[6] Similarly, the unique "double-beam" chevrons of the mid-caudal vertebrae (that give rise to its name) could have helped distribute the load of the animal when balancing tripodally on the tail, and even the large thumb claws might have helped support a rearing *Diplodocus* by anchoring onto tree trunks.[278]

Although Barrett and Upchurch[394] suggested that the smaller head of *Diplodocus* would have given it a more precise bite than that of large-headed contemporaries like the early macronarian *Camarasaurus*, John Whitlock[396] pointed to the overall square shape of the skull of *Diplodocus* as in fact being that of an unselective bulk feeder. He noted that this also tended to correlate with the animal being a low browser and noted that microwear patterns on tooth enamel (specifically, the appearance of pits) found *Diplodocus* to have an intermediate number between those of high feeders like *Brachiosaurus* and low feeders like the relatively short-necked diplodocoid *Nigersaurus*. Other microwear marks—fine, subparallel striations—were similar to those of *Nigersaurus* and pointed to feeding on low and softer stems. In addition, analysis of the histology of *Diplodocus* teeth suggests that they were not suited to the stresses of a branch-stripping type of feeding, and that a more typical cropping action would have been the norm.[397]

In short, *Diplodocus* at least seems more than capable of feeding both high and low. It likely was able to shift effectively between these very different modes of feeding and was not reliant on a single one. Such a pattern is echoed in giraffes, which, while preferentially feeding high up, are often forced into lower feeding through lack of suitable browse.[398] Other sauropods also show suggestions of high feeding: the early sauropod *Spinophorosaurus*, for example, shows a very front-high body posture, which would place the head in a high position with an upwardly titled pelvis—a key innovation[399] that would make anything but high foraging difficult. Doubtless there was variation within the sauropods (and many would have changed as they grew), but it is still unclear for many taxa whether they were fundamentally high or low feeders.

CHAPTER 9

Summary

Throughout this book it should be clear that understanding and interpreting the behavior of dinosaurs is profoundly difficult, given the limitations of the fossil record and the plasticity and diversity of behavior in animals. It is all but impossible to know the exact circumstances surrounding the moment that an animal died or its traces were laid down, and the issues that may have affected its behavior are unknown. Those tracks showing a theropod moving swiftly alongside a much larger sauropod[400] could be the result of a predation attempt, but the theropod might have been chasing off an animal that got too close to its young, or driving away a competitor from a water hole; the theropod itself might have been a herbivore rather than a carnivore, and moving in a mixed-species herd with the sauopod; or both might have been running from a larger theropod whose tracks were not preserved. Any of these are possibilities, but even if we are correct that this was an attempt to hunt down prey, it might have been a very unusual situation: a naïve theropod trying to tackle prey way out of its league (which would not normally happen); or a sauropod that was aging, injured, or ill and therefore vulnerable, or unusual in having gotten lost from its herd; or the carnivore suffering from parasites or dehydration that was blindly attacking any other animal it encountered. Any or all of these might also be possible; it is so easy to interpret these tracks as evidence that "medium-sized theropods hunted larger sauropods" and even be technically correct, while taking an exceptional and unusual event and treating it as normal, and worse, extrapolating it across huge numbers of near relatives. What one individual of a species does in a given situation may not be what even a sibling would do in the same circumstances, let alone a different species that lived 10 million years later on another continent and in a different ecosystem.

Thus, while singular fossils of trackways of animals moving together or animals apparently locked in mortal combat are extremely
interesting and engaging, they may be actively misleading as to the
lives of the animals that left them. While this means that almost any
interpretation of dinosaur behavior is going to be left with a lot of
uncertainty, I think it is greatly preferable to be left with uncertainty
and limited to generalities than to move forward with what may be
incorrect assumptions and inferences, and risk building hypotheses
on very unstable foundations. Even so, much of the scientific literature
tends toward a confidence in interpreting dinosaurian behaviors that
probably should not be there, and a failure to recognize alternate
possibilities and the inherent uncertainty of interpreting ancient
behaviors is a detriment to the field.

That said, the advances that have been made are obviously enormous. It would indeed be a sorry state of affairs if the study of dinosaurian
ethology was still taking a Victorian approach, but it has improved
massively in recent years. Early statements about the behavior of
animals based on a single feature, or with considerable guesswork,
extrapolation, and imagination, have been replaced by as much rigor
as is possible within the limitations described above, and by the approach of trying to understand dinosaurs as real and living animals,
not near-mythical and unknowable oddities. Moreover, a much better
understanding of the relationships of dinosaurs to other reptiles and
birds, and a better knowledge of their biology, gives a much more robust evolutionary framework for working out what behaviors might
be possible and which selective pressures might operate under various
conditions. This is especially true when considering that we have far
better understanding of the biological bases of behavior in general,
and the behavior of reptiles in particular as a starting point.

Perhaps most obvious, but no less important, is the sheer number
of fossils that are now available for study (figure 9.1). Two centuries of
excavations mean that rare events have a chance of being captured,
and what were previously anomalies (like monospecific bonebeds,
gut contents, or bite traces) have become commonplace and can be
accepted as normal. In the context of the inherent uncertainty of
studying behavior in the fossil record, such repeat discoveries are

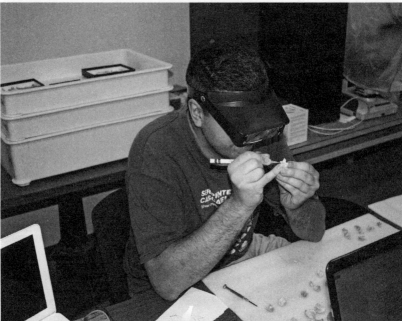

FIGURE 9.1 Dinosaur research continues with new discoveries in the field, and research in museums, universities, and institutes around the world. Photographs courtesy of Corwin Sullivan.

especially important for determining what may be typical for a species or clade, and make for a robust starting point for testing hypotheses and assessing other information.

This enriched evidence base has been coupled with adding many more lines of data to the assessments of dinosaur behavior. Evidence like stable isotope data from teeth, biomechanics and functional morphological assessments of skeletal anatomy, and environmental data from the rock record all come together to build up (in some cases at least) very strong arguments for what certain animals are doing. We have moved on from "sauropods must have fed high in trees because they had long necks" to being able to look at the flexibility and support of the neck from a functional perspective; head position from the orientation of the inner ear; tooth wear and scratches on the enamel; stomach contents from fossils; palynological analysis of the local flora, and more to build up a comprehensive picture of what animals fed on and how. These new analyses and techniques (and increasing interest in the field from an ever-growing number of researchers) mean that our ability to work out such things as the details of diets has the potential for exponential improvement.

This new data may also reach some kind of tipping point, for a handful of species at least. So many aspects of behavior are interlinked, and issues such as clutch sizes, sexual displays, population size and density, local environment and resource availability can all have major impacts on each other. Data on only one or two of these can offer only the most limited glimpses of what animals might be doing, but convincing data on many or all of them would allow the buildup of a real picture of the behavior of an animal, and lead to a confident assertion about, say, a polygamous mating system, even when direct evidence for that will never be available. Integrating the new information, and specifically exploring areas of dinosaur paleobiology that might reveal complementary data, is also important.

What next? Certainly we can expect more specimens, at least, and more researchers, and ever newer techniques and technologies that can applied by those scientists to those specimens. One area in which I hope for further improvement would be the engagement of more ethologists in the realm of paleontology. This has been a pattern in recent years, as ecologists who normally work on extant systems have brought their expertise into collaborations with paleontologists. The same shift would be most welcome with experts in behavior, and the ability to look at the origins of behaviors in lineages like the theropods on the way to birds would surely benefit ornithological studies as much as it would paleontological, to the enrichment of all concerned. Similarly, the refinement of data and improving sources on the behavior of living animals will provide a much more concrete basis for making inferences and testing hypotheses about dinosaurs (figure 9.2). Where, for example, the diet of living taxa is uncertain, then using this as a model for interpreting dinosaurian feeding behaviors based on shared anatomy becomes problematic. Improvements here (including ethologists and ecologists explicitly collecting the kinds of data that paleontologists can use) will also be of enormous benefit.

For all this, the behavior of extinct animals remains hard to study, and we may never find evidence of even some basic behaviors (or we may have evidence and be unable to interpret it, or unable to rule out alternatives). There will always be major gaps in our understanding of the day-to-day lives of dinosaurs, and even of major aspects of their behavioral ecology, that seem near impossible to fill with any confidence (how long did they sleep? Did they hold territories?), but improvements in our knowledge will come. Although we are, and will always be, missing some basic information, what is perhaps more intriguing (and even infuriating) is our inability to see any of the quirks of behavior that would show up more rarely. The partnerships between species that hunt together or groom each other; tool-using behaviors or unusual hunting strategies; misdirected protection by mothers of another species' offspring; incorrect matings between species. These are the rarities and oddities that show the real depth and breadth of behaviors, and the possibilities for what animals can do under unusual circumstances, will be impossible to determine in the fossil record.

So even while I have generally cautioned here against over-extrapolation from limited data and have counseled embracing uncertainty, it is perhaps also worth remembering all the incredible

FIGURE 9.2 The study of live animals is a major source of data on dinosaur biology. Here a young Nile crocodile (*Crocodylus niloticus*) (top) and an elegant crested tinamou (*Eudromia elegans*) (bottom) are filmed with X-ray video (XROMM) to study how their bones and joints operate during locomotion, to help determine the way dinosaurs likely moved. Photographs courtesy of John Hutchinson.

FIGURE 9.3 While paleontologists are limited to (mostly) skeletal remains of the dinosaurs, we remain well placed to explore the behaviors of these incredible animals. Here, a dramatic mount of a pair of *Dryptosaurus* at the New Jersey State Museum shows the very kinds of actions we cannot directly observe, but can infer with careful study. Photograph courtesy of Skye McDavid.

things that living animals do, and recognizing that the sum total of the behaviors of dinosaurs across 1,500 species and 170 million years were undoubtedly much more varied and much more complex and interesting than we can ever know. So, whatever inferences and reconstructions we can make (figure 9.3), the search for this information will, no doubt, be most rewarding.

Dinosaur Clades

The first chapter in this book contains only the barest of introductions to the three major clades of dinosaurs, and some knowledge of the dinosaurian "families" and their biology is an important context for understanding their behavior. Also important are their general relationships to each other, since, although behavior is highly plastic and varies enormously both within and between species, animals with a recent shared ancestry, body plan, and diet, living in similar ecosystems with similar conditions, are likely to share similar behaviors. Thus information about their essential relationships (to other archosaurs and within the dinosaurs—figure 10.1) is an important consideration when attempting to extrapolate from limited data on one species to other animals.

The list and brief descriptions given here are not presented as a phylogeny, but the major groups are in the approximate order of branching from earliest to latest, and the indentation of names is used to indicate that they belong in the group above. Dinosaur phylogeny is, inevitably, in a state of some flux, and while the outline given here is broadly stable, some relationships between clades remain unresolved, and these may move (or disappear) in the future. Also, not all of the various dinosaur clades are covered in this list; there are numerous small groups that are represented by only a few animals and are poorly known or have been little studied in terms of their behavior, and so would be extraneous to include here, since they will not feature anywhere else in the book.

Only the most basic information is given here about the biology and appearance of each group, and many of them contain considerable variation. For example, the tyrannosaurs include at least 30 species, ranged in size from around 3 m to over 12 m long and perhaps from 100 kg to >7 tons. While all were predators, they varied in the size

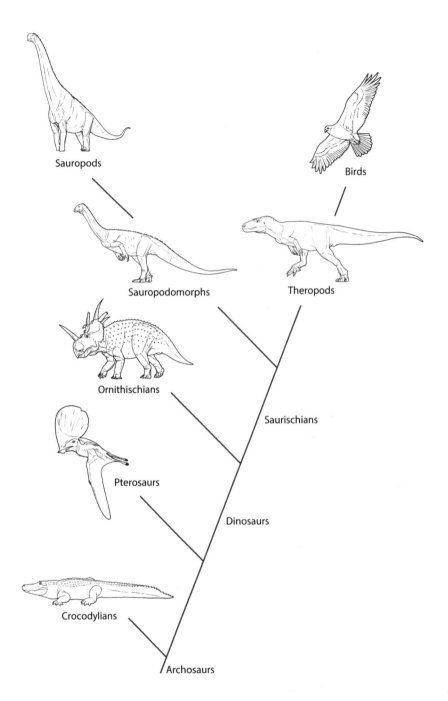

FIGURE 10.1 The major lineages of the archosaurs, both living and extinct: Theropoda (top, *Megalosaurus*; far right, the bird *Pandion*), Sauropodomorpha (*Giraffatitan*), and Ornithischia (center, *Stegosaurus* and right, *Styracosaurus*); Crocodilia (*Alligator*) and Pterosauria (*Caiuajara*). Artwork by Gabriel Ugueto.

and shape of their skulls and teeth, and early forms had long arms with three fingers, unlike the famously small-armed, two-fingered giants. They spanned around 100 million years of time and lived in North America, Europe, and Asia, at least, though they may also have been in Australia and could include a contingent of primarily South American animals. Capturing their diversity, disparity, biogeography, and evolutionary history is therefore impossible in a few lines, but the basic information below should be sufficient to cover the essential information about these clades as a primer. In short, readers should be able to refer to this section easily and at least identify the types of dinosaurs being discussed if they are not already familiar with them. Often in the text I will use anglicized forms of these names (e.g., ankylosaurid for Ankylosauridae), but they can be readily recognized and treated synonymously and similarly. I will not specify that genera who are the exemplars of clades are members (it's rather redundant to say that *Mamenchisaurus* is a member of the Mamenchisauridae).

Theropoda (Figure 10.2)

Herrerasauria This short-lived group of Late Triassic carnivores are of uncertain position and may sit outside of the theropods as an early branch of Dinosauria. They were mostly small-bodied animals (2–3 m long), though with relatively large heads for their size, unlike many other early dinosaur groups.

Coelophysoidea These small dinosaurs, known primarily from North America and Africa, were notable for their slender build, long necks, and long, thin skulls with numerous small teeth. Some have been found in tremendous numbers together.

Ceratosauria This group includes some of the earliest large-bodied (>5 m) dinosaurian predators. Some early members of this group evolved to become the first theropodan herbivores, and possessed with beaks; others were also unusual for theropods in having some limited bony armor in the skin. Many bore bony crests on the head.

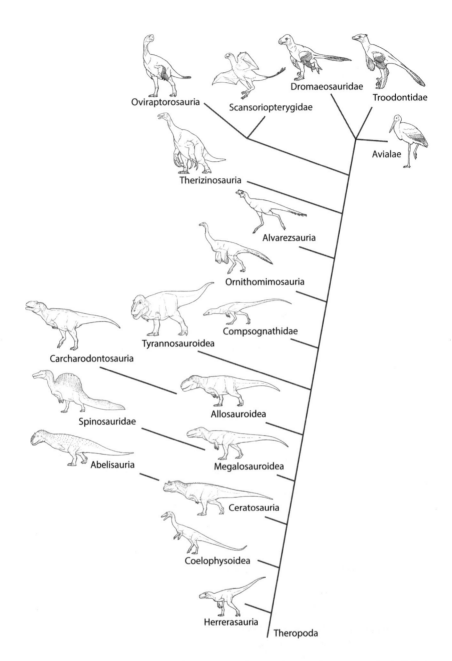

FIGURE 10.2 The major lineages of the Theropoda. *Gnathovorax* (Herrerasauria), *Coelophysis* (Coelophysoidea), *Ceratosaurus* (Ceratosauria), *Majungasaurus* (Abelisauria), *Torvosaurus* (Megalosauroidea), *Spinosaurus* (Spinosauridae), *Yangchuanosaurus* (Allosauroidea), *Meraxes* (Carcharodontosauria), *Tyrannosaurus* (Tyrannosauroidea), *Compsognathus* (Compsognathidae), *Gallimimus* (Ornithomimosauria), *Linhenykus* (Alvarezsauria), *Therizinosaurus* (Therizinosauria), *Gigantoraptor* (Oviraptorosauria), *Yi* (Scansoriopterygidae), *Deinonychus* (Dromaeosauridae), *Hesperornithoides* (Troodontidae), and a modern stork, *Ciconia* (to represent Aves). Artwork by Gabriel Ugueto.

Abelisauria These Cretaceous ceratosaurs were predominantly from the southern hemisphere. They are notable for bearing large horns or bosses on their deep and narrow skulls, and for having extremely reduced forelimbs that retained four very short fingers.

Megalosauroidea A diverse group of mostly Jurassic carnivorous theropods, their remains are often fragmentary; thus they are not especially well known and little studied.

Spinosauridae A Cretaceous offshoot of the megalosauroids, these are famous for their crocodile-like snouts, large arms, at least partially piscivorous diet, and elongate spines on the dorsal vertebrae, which, in some, produce an enormous sail along the back. Some of these were among the largest known theropods.

Allosauroidea Perhaps the dominant group of Jurassic predators, these were mostly large animals (>8 m) with both large teeth and large claws on strong arms. They had a wide distribution and are found in very large numbers at some sites.

Carcharodontosauria These members of the allosauroids included some very large South American, African, and Asian species that had narrow skulls. In the northern hemisphere they were replaced by the large tyrannosaurs in the later part of the Cretaceous.

Tyrannosauroidea Famous for the later members of this group that include the largest terrestrial predators of all time, derived tyrannosaurs were robust and have large heads with reduced numbers of large teeth and very small forelimbs. Earlier versions were smaller and less specialized, and all members had at least some form of small crests on the head. At least some tyrannosaurs were feathered, as were all subsequently listed theropod groups.

Compsognathidae An enigmatic group of small, feathered predatory animals from Europe and Asia are known from juvenile specimens. This has led to the recent suggestion that they may actually be young members of other early theropod groups.

Ornithomimosauria These predominantly northern hemisphere animals were mostly long-legged and lightly built, with a generally ostrich-like appearance, having small heads on a long neck and long, thin arms with reduced claws. Many were beaked herbivores, but at least some early species had many small teeth.

Alvarezsauria An unusual group of mostly very small animals, including the smallest non-bird dinosaurs known, which as adults were about the size of crows. They had small, beaked heads, and this, coupled with very small, but very strong arms bearing a single giant claw, has led to their being considered insect-eating specialists. They were present in Europe, Asia, and North and South America.

Therizinosauria Another group of herbivorous northern hemisphere theropods, these are notable for their long necks, long arms, and large, or even sometimes giant, claws. These Cretaceous animals are known primarily from North America and Asia and were mostly rare, though a couple of species are known from numerous individuals. They were generally robust and stocky animals with a beak in front of numerous small teeth.

Oviraptorosauria This clade includes some animals that were likely herbivorous and others that may have been omnivores. They present a mixture of toothed animals, those with teeth and beaks, and those with just beaks. These are the first group here to have pennaceous feathers like those of modern birds, and many had a very birdlike appearance, with long feathers on the arms and a fan on the end of the tail.

    Scansoriopterygidae This group of tiny animals were arboreal and combined feathers with a flying squirrel–like wing membrane to allow them to glide. They are currently only known from a handful of specimens from China; their biology and exact relationships are not well understood, and their affinity here as an unusual offshoot of early oviraptorosaurs is uncertain.

Dromaeosauridae These predators were mostly small and include a number of species that could at least glide, and perhaps were capable of powered flight. These animals are notable for their long leg and arm feathers (also seen in the subsequent groups) and the large claw of the foot that was retracted and kept free of the ground, leaving them walking on two toes.

Troodontidae An enigmatic group of small birdlike animals that have unusual teeth with giant serrations on them, leading to suggestions of both carnivorous and herbivorous diets. Some were possibly arboreal and also capable of limited flight.

Avialae These dinosaurs are the closest to—and also include—modern birds. A number of forms have "dinosaurian" traits like large hand claws, teeth, and long bony tails, but numerous Cretaceous groups would appear indistinguishable from modern birds at first glance.

Sauropodomorpha (Figure 10.3)

Early sauropodomorphs ("prosauropods") These dinosaurs were small bipeds and very similar to early theropods. They had relatively long necks and small heads, and the earliest were most likely omnivorous, or possibly even carnivorous. These and their descendants show limited development of pneumatic systems in the skeleton. All later animals would be herbivores.

Plateosauria This prosauropod group were the largest of the Triassic dinosaurs. Still predominantly bipedal (though some were quadrupeds when young), they could reach over a ton in weight and were specialized herbivores. The small head on a long neck was a clear sign of what would come later, and the large chest allowed for a long gut to digest tough plants. The hands generally bore large claws.

Sauropoda Technically nested within Plateosauria, the sauropods are instantly recognizable for their huge size and very long necks bearing small heads. These were all quadrupeds, and often had simple teeth, large claws on their thumbs, and long tails. They were also highly

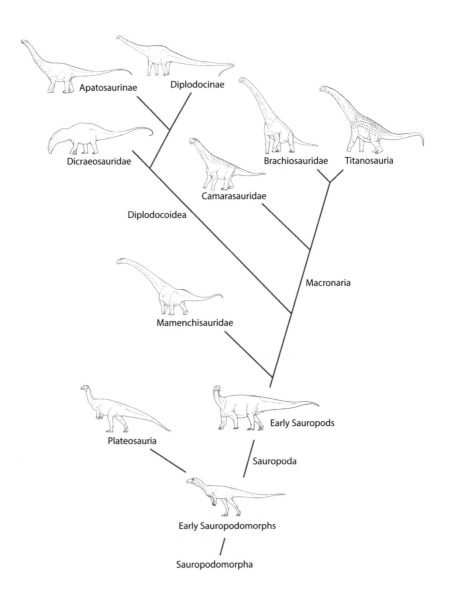

FIGURE 10.3 The major lineages of the Sauropodomorpha. *Panphagia* (early sauropodomorphs), *Plateosaurus* (Plateosauria), *Ingentia* (early sauropods), *Mamenchisaurus* (Mamenchisauridae), *Amargasaurus* (Dicraeosauridae), *Diplodocus* (Diplodocinae), *Apatosaurus* (Apatosaurinae), *Camarasaurus* (Camarasauridae), *Brachiosaurus* (Brachiosauridae), and *Alamosaurus* (Titanosauria). Artwork by Gabriel Ugueto.

pneumatic with extensive air sacs that invaded the spine especially, to reduce their weight.

Early sauropods These animals were somewhat "generic" in shape, and while they have the bauplan of sauropods, they lack the extremes of neck length, tail length, or overall size of later clades. The Chinese *Shunosaurus* and some near relatives are notable, though, for the presence of a bony tail club.

Mamenchisauridae These Jurassic Asian dinosaurs pushed the length of the neck to the extreme. Despite the presence of a long tail, the neck could make up half the total length of the animal, giving it an extraordinary reach. This was supported by very long cervical ribs, which could be 2 m long and overlapped one another to support the neck.

Diplodocoidea This major group of sauropods includes numerous famous Jurassic animals (though they were also present in the Cretaceous), and while many representatives are from North America, they were also present in South America, Europe, Asia, and Africa.

Dicraeosauridae Known from only a few species, this group nonetheless has representatives across multiple continents and was very widespread. In addition to being among the smallest of the sauropods, they are notable for two major features of the neck. First off, this was so short it could barely reach the ground; and second, it often had an array of spikes sticking up from it.

Apatosaurinae These dinosaurs had necks that were long but also very broad, and boss-like expansions on their undersides, on a body that was generally more heavyset than that of the other diplodocids.

Diplodocinae This group are perhaps the best known of the sauropods and are characterized by their often extremely long and whiplike tails, which have numerous short, rod-like bones in them. These may have been high-browsing animals, stripping the leaves from branches with their teeth.

Macronaria This derived group of sauropods includes both the tallest and the heaviest of the sauropods. Named for their large nostrils, these animals often have a dome-like expansion to the snout.

Camarasauridae Exemplified by *Camarasaurus*, these are perhaps the most common Jurassic sauropod genus (and known from a huge number of finds). The camarasaurs had a relatively short and tall skull with large teeth and a relatively short neck on a robust body.

Brachiosauridae These large sauropods had rather long forelimbs, and this gave them the appearance of a lowered back end. Unusually, they also held their necks in a more vertical posture than most other sauropods, and likely fed high into trees.

Titanosauria This diverse group of Cretaceous sauropods includes the largest known dinosaurs. The titanosaurs were especially common in southern continents and include a wide variety of forms, including some with bits of dermal armor and many who lost their manual digits and walked on their metacarpals.

Ornithischia (Figure 10.4)

Heterodontosauridae These are the smallest of the ornithischians and are known from finds in Africa, Asia, and North America. These were long-lasting and are represented in both the Jurassic and Cretaceous. At least some were covered in feather-like filaments, and all were small bipeds with a pair of large fangs at the front of the jaw, though they were predominantly herbivorous.

Thyreophora This is a major group of quadrupedal dinosaurs that all have some form of bony armor or spikes and plates across their bodies. They had near global distribution and had members from the early Jurassic to the end of the Cretaceous.

Stegosauria These animals had two rows of plates and spikes along their bodies, usually with plates on the front half and spikes on the rear, including the tail tip. Most also had a pair of large spines on the shoulders. These were low browsers, though a couple of taxa had rather longer necks and may have been higher feeders.

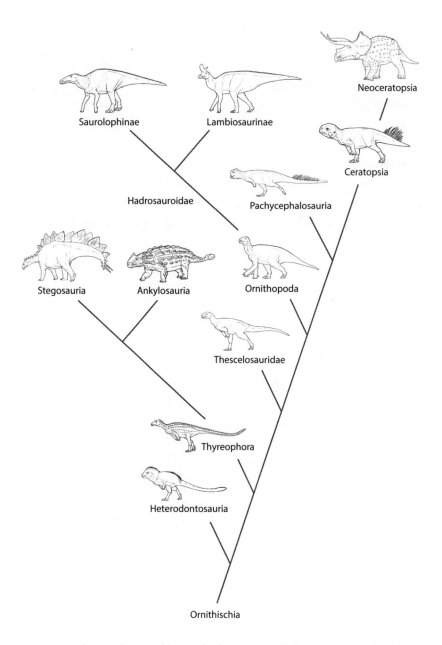

FIGURE 10.4 The major lineages of the Ornithischia represented by key genera. *Tianyulong* (Heterodontosauria), *Scutellosaurus* (Thyreophora), *Stegosaurus* (Stegosauria), *Ankylosaurus* (Ankylosauria), *Parksosaurus* (Thescelosauridae), *Iguanodon* (Ornithopoda), *Edmontosaurus* (Saurolophinae), *Lambeosaurus* (Lambeosaurinae), *Prenocephale* (Pachycephalosauria), *Psittacosaurus* (Ceratopsia), *Triceratops* (Neoceratopsia). Artwork by Gabriel Ugueto.

Ankylosauria The famous armored dinosaurs whose backs and flanks were almost entirely covered in in armored bumps and spikes, with skulls entirely encased in bone. Within this group there is a major division between the ankylosaurids, which had a large bony tail club, and the nodosaurids, which lacked the club but had large shoulder spikes.

Thescelosauridae A group of mostly small to mid-size, facultatively bipedal animals that lack any of the anomalous features that characterize other ornithischian clades. Even so, the group includes the odd *Oryctodromeus* which may have excavated, and lived in, burrows.

Ornithopoda Another major division, these large Cretaceous animals, predominantly from the northern continents, had specialized jaws with numerous rows of continually growing teeth to process tough plants prior to swallowing. Their relatively short forelimbs, compared to the hindlimbs of many, had led to the idea that they were bipeds, but they are now regarded as being fundamentally quadrupedal.

 Early ornithopods ("Iguanodontids") Most of these animals, *Iguanodon* and kin, are well known for the possession of spike-like thumbs, but they also had a partially opposable fifth finger on the hand to give them some grasping ability. They were very common in Europe but also were in Asia and Africa.

 Hadrosauridae This group from the Late Cretaceous were diverse and were the dominant large herbivores in the northern hemisphere. Often (erroneously) called "duck-billed" for their broad mouths, they actually had sharp and ventrally directed beaks. Two major groups are known: the lambeosaurines, which had an array of hollow bony expansions to the skull, and the saurolophines, which generally lacked bony crests.

Pachycephalosauria A bipedal group that included both small and large animals. Their fossil record is mostly very poor except for their skulls, as they had a tremendously thickened and expanded dome of bone on their heads. The function of this has been the subject of enormous debate, but these were likely used to ram one another.

Ceratopsia This is a great group of horned dinosaurs that lived across Asia and North America, with a handful outside those continents. Early forms from the Late Jurassic and Early Cretaceous were small and bipedal. Notably, the early animals lacked any real frill at the back of the head or horns on the face, but instead had a pair of bosses on the cheeks. At least one had feather-like filaments.

Neoceratopsia These Late Cretaceous animals were much larger than their forebears and were all quadrupedal. All had some kind of large frill on the back of the head, and most had some combination of horns on their faces as well. They were the most common herbivores of the time, after the hadrosaurs.

REFERENCES

1 Buckland, W. 1824. XXI.—Notice on the *Megalosaurus* or great Fossil Lizard of Stonesfield. *Transactions of the Geological Society of London*, 2:390–396.

2 Mantell, G.A. 1825. Notice on the *Iguanodon*, a newly discovered fossil reptile, from the sandstone of Tilgate forest, in Sussex. *Philosophical Transactions of the Royal Society*, 115:179–186.

3 Owen, R. 1842. Report on British fossil reptiles, part II. *Report for the British Association for the Advancement of Science, Plymouth, 1841*, 60–294.

4 Weishampel, D.B., Barrett, P.M., Coria, R.A., Le Loeuff, J., Xing, X., Xijin, Z., Sahni, A., Gomani, E.M., and Noto, C.R. 2004. Dinosaur distribution. In *The Dinosauria*, 2nd ed., 517–606. University of California Press, Berkeley.

5 Fastovsky, D.E., and Smith, J.B. 2004. Dinosaur Paleoecology. In *The Dinosauria*, 2nd ed., 614–626. University of California Press, Berkeley.

6 Hallett, M., and Wedel, M.J. 2016. *The Sauropod Dinosaurs: Life in the Age of Giants*. John Hopkins University Press, Baltimore.

7 Xu, X., Zhou, Z., Wang, X., Kuang, X., Zhang, F., and Du, X. 2003. Four-winged dinosaurs from China. *Nature*, 421:335–340.

8 Bell, P.R., Fanti, F., Currie, P.J., and Arbour, V.M. 2014. A mummified duck-billed dinosaur with a soft-tissue cock's comb. *Current Biology*, 24:70–75.

9 Zhang, F., Kearns, S.L., Orr, P.J., Benton, M.J., Zhou, Z., Johnson, D., Xu, X., and Wang, X. 2010. Fossilized melanosomes and the colour of Cretaceous dinosaurs and birds. *Nature*, 463:1075–1078.

10 Benton, M.J. 2014. *Vertebrate Palaeontology*. John Wiley & Sons.

11 Nesbitt, S.J., Barrett, P.M., Werning, S., Sidor, C.A., and Charig, A.J. 2013. The oldest dinosaur? A Middle Triassic dinosauriform from Tanzania. *Biology Letters*, 9:20120949.

12 Benson, R.B. 2018. Dinosaur macroevolution and macroecology. *Annual Review of Ecology, Evolution, and Systematics*, 49:379–408.

13 Baron, M.G., Norman, D.B., and Barrett, P.M. 2017. A new hypothesis of dinosaur relationships and early dinosaur evolution. *Nature*, 543:501–506.

14 Müller, R.T., and Garcia, M.S. 2020. A paraphyletic 'Silesauridae' as an alternative hypothesis for the initial radiation of ornithischian dinosaurs. *Biology Letters*, 16:20200417.

15 Hendrickx, C., Hartman, S.A., and Mateus, O. 2015. An overview of non-avian theropod discoveries and classification. *PalArch's Journal of Vertebrate Palaeontology*, 12:1–73.

16 Barrett, P.M. 2014. Paleobiology of herbivorous dinosaurs. *Annual Review of Earth and Planetary Sciences*, 42:207–230.

17 Xu, X., Zhou, Z., Dudley, R., Mackem, S., Chuong, C.M., Erickson, G.M., and Varricchio, D.J. 2014. An integrative approach to understanding bird origins. *Science*, 346:1253293.

18 Taylor, M.P., and Wedel, M.J. 2013. Why sauropods had long necks; and why giraffes have short necks. *PeerJ*, 1:e36.

19 Godefroit, P., Sinitsa, S.M., Cincotta, A., McNamara, M.E., Reshetova, S.A., and Dhouailly, D. 2020. Integumentary structures in *Kulindadromeus zabaikalicus*, a basal neornithischian dinosaur from the Jurassic of Siberia. In: *The Evolution of Feathers*, 47–65. Springer, Cham.

20 Langer, M.C., de Oliveira Martins, N., Manzig, P.C., de Souza Ferreira, G., de Almeida Marsola, J.C., Fortes, E., Lima, R., Sant'ana, L.C.F., da Silva Vidal, L., da Silva Lorençato, R.H., and Ezcurra, M.D. 2019. A new desert-dwelling dinosaur (Theropoda, Noasaurinae) from the Cretaceous of south Brazil. *Scientific Reports*, 9:1–31.

21 Tykoski, R.S., Fiorillo, A.R., and Chiba, K. 2019. New data and diagnosis for the Arctic ceratopsid dinosaur *Pachyrhinosaurus perotorum*. *Journal of Systematic Palaeontology*, 17:1397–1416.

22 Fricke, H.C., Rogers, R.R., and Gates, T.A. 2009. Hadrosaurid migration: inferences based on stable isotope comparisons among Late Cretaceous dinosaur localities. *Paleobiology*, 35:270–288.

23 Barrett, P.M., and Rayfield, E.J. 2006. Ecological and evolutionary implications of dinosaur feeding behaviour. *Trends in Ecology & Evolution*, 21:217–224.

24 Chin, K. 2012. What did dinosaurs eat: coprolites and other direct evidence of dinosaur diet. In: *The Complete Dinosaur*, 2nd ed., 589–602. Indiana University Press, Bloomington.

25 Maidment, S.C., and Barrett, P.M. 2012. Osteological correlates for quadrupedality in ornithischian dinosaurs. *Acta Palaeontologica Polonica*, 59:53–70.

26 Henderson, D.M. 2004. Tipsy punters: sauropod dinosaur pneumaticity, buoyancy and aquatic habits. *Proceedings of the Royal Society of London B: Biological Sciences*, 271:S180–S183.

27 Henderson, D.M. 2018. A buoyancy, balance and stability challenge to the hypothesis of a semi-aquatic *Spinosaurus* Stromer, 1915 (Dinosauria: Theropoda). *PeerJ*, 6:e5409.

28 Naish, D. 2000. Theropod dinosaurs in the trees: a historical review of arboreal habits amongst nonavian theropods. *Archaeopteryx*, 18:35–41.

29 Fearon, J.L., and Varricchio, D.J. 2015. Morphometric analysis of the forelimb and pectoral girdle of the Cretaceous ornithopod dinosaur *Oryctodromeus cubicularis* and implications for digging. *Journal of Vertebrate Paleontology*, 35:e936555.

30 Pei, R., Pittman, M., Goloboff, P.A., Dececchi, T.A., Habib, M.B., Kaye, T.G., Larsson, H.C., Norell, M.A., Brusatte, S.L., and Xu, X. 2020. Potential for powered flight neared by most close avialan relatives, but few crossed its thresholds. *Current Biology*, 30(20):4033–4046.

31 Qi, Z., Barrett, P.M., and Eberth, D.A. 2007. Social behaviour and mass mortality in the basal ceratopsian dinosaur *Psittacosaurus* (Early Cretaceous, People's Republic of China). *Palaeontology,* 50:1023–1029.

32 Lockley, M., Schulp, A.S., Meyer, C.A., Leonardi, G., and Mamani, D.K. 2002. Titanosaurid trackways from the Upper Cretaceous of Bolivia: evidence for large manus, wide-gauge locomotion and gregarious behaviour. *Cretaceous Research,* 23:383–400.

33 Hone, D.W.E., and Mallon, J.C. In press. Dinosaurs in groups—social behaviours, interactions and communication. In: *The Complete Dinosaur*, 3rd ed. Indiana University Press, Bloomington.

34 Hone, D.W., Naish, D., and Cuthill, I.C. 2012. Does mutual sexual selection explain the evolution of head crests in pterosaurs and dinosaurs?. *Lethaia*, 45:139–156.

35 Lockley, M.G., McCrea, R.T., Buckley, L.G., Lim, J.D., Matthews, N.A., Breithaupt, B.H., Houck, K.H., Gierliński, G.D., Surmik, D., Kim, K.S., and Xing, L. 2016. Theropod courtship: large scale physical evidence of display arenas and avian-like scrape ceremony behaviour by Cretaceous dinosaurs. *Scientific Reports*, 6:18952.

36 Farke, A.A., Wolff, E.D., and Tanke, D.H. 2009. Evidence of combat in *Triceratops*. *PLoS One,* 4:e4252.

37 Brown, C.M., Currie, P.J., and Therrien, F. 2021. Intraspecific facial bite marks in tyrannosaurids provide insight into sexual maturity and evolution of bird-like intersexual display. *Paleobiology*, 48:12–43.

38 Happ, J. 2008. An analysis of predator–prey behavior in a head-to-head encounter between *Tyrannosaurus rex* and *Triceratops*. In: Tyrannosaurus rex, *the Tyrant King*, 352–370. Indiana University Press, Bloomington.

39 Senter, P. and Juengst, S.L. 2016. Record-breaking pain: the largest number and variety of forelimb bone maladies in a theropod dinosaur. *PLoS One*, 11:e0149140.

40 Hamm, C.A., Hampe, O., Schwarz, D., Witzmann, F., Makovicky, P.J., Brochu, C.A., Reiter, R., and Asbach, P. 2020. A comprehensive diagnostic approach combining phylogenetic disease bracketing and CT imaging reveals osteomyelitis in a *Tyrannosaurus rex*. *Scientific Reports*, 10:1–16.

41 Wolff, E.D., Salisbury, S.W., Horner, J.R., and Varricchio, D.J. 2009. Common avian infection plagued the tyrant dinosaurs. *PLoS One*, 4:e7288.

42 Grellet-Tinner, G., Chiappe, L., Norell, M., and Bottjer, D. 2006. Dinosaur eggs and nesting behaviors: a paleobiological investigation. *Palaeogeography, Palaeoclimatology, Palaeoecology*, 232:294–321.

43 Norell, M.A., Clark, J.M., Chiappe, L.M., and Dashzeveg, D. 1995. A nesting dinosaur. *Nature*, 378(6559):774–776.

44 Erickson, G.M., Zelenitsky, D.K., Kay, D.I., and Norell, M.A. 2017. Dinosaur incubation periods directly determined from growth-line counts in embryonic teeth show reptilian-grade development. *Proceedings of the National Academy of Sciences*, 114:540–545.

45 Werning, S. 2012. The ontogenetic osteohistology of *Tenontosaurus tilletti*. *PLoS One*, 7:e33539.

46 Reisz, R.R., Scott, D., Sues, H.D., Evans, D.C., and Raath, M.A. 2005. Embryos of an Early Jurassic prosauropod dinosaur and their evolutionary significance. *Science*, 309:761–764.

47 Wang, S., Stiegler, J., Amiot, R., Wang, X., Du, G.H., Clark, J.M., and Xu, X. 2017. Extreme ontogenetic changes in a ceratosaurian theropod. *Current Biology*, 27:144–148.

48 Carr, T.D. 2020. A high-resolution growth series of *Tyrannosaurus rex* obtained from multiple lines of evidence. *PeerJ*, 8:e9192.

49 Holtz, T.R. 2021. Theropod guild structure and the tyrannosaurid niche assimilation hypothesis: implications for predatory dinosaur macroecology and ontogeny in later Late Cretaceous Asiamerica. *Canadian Journal of Earth Sciences*, 99:778–795.

50 O'Brien, D.M., Allen, C.E., Van Kleeck, M.J., Hone, D., Knell, R., Knapp, A., Christiansen, S., and Emlen, D.J. 2018. On the evolution of extreme structures: static scaling and the function of sexually selected signals. *Animal Behaviour*, 144:95–108.

51 Chapelle, K.E., Botha, J., and Choiniere, J.N. 2021. Extreme growth plasticity in the early branching sauropodomorph *Massospondylus carinatus*. *Biology Letters*, 17:20200843.

52 Barta, D.E., Griffin, C.T., and Norell, M.A. 2022. Osteohistology of a Triassic dinosaur population reveals highly variable growth trajectories typified early dinosaur ontogeny. *Scientific Reports*, 12:1–14.

53 Griffin, C.T., Stocker, M.R., Colleary, C., Stefanic, C.M., Lessner, E.J., Riegler, M., Formoso, K., Koeller, K., and Nesbitt, S.J. 2021. Assessing ontogenetic maturity in extinct saurian reptiles. *Biological Reviews*, 96:470–525.

54 Hone, D.W., Farke, A.A., and Wedel, M.J. 2016. Ontogeny and the fossil record: what, if anything, is an adult dinosaur?. *Biology Letters*, 12:20150947.

55 Brasier, M.D., Norman, D.B., Liu, A.G., Cotton, L.J., Hiscocks, J.E., Garwood, R.J., Antcliffe, J.B., and Wacey, D. 2017. Remarkable preservation of brain tissues in an Early Cretaceous iguanodontian dinosaur. *Geological Society, London, Special Publications*, 448:383–398.

56 Balanoff, A.M., and Bever, G.S. 2020. The role of endocasts in the study of brain evolution. In: *Evolutionary Neuroscience*, 29–49. Academic Press.

57 Witmer, L.M., Ridgely, R.C., Dufeau, D.L., and Semones, M.C. 2008. Using CT to peer into the past: 3D visualization of the brain and ear regions of birds, crocodiles, and nonavian dinosaurs. In: *Anatomical Imaging*, 67–87. Springer, Tokyo.

58 Woodruff, D.C., Naish, D., and Dunning, J. 2021. Photoluminescent visual displays: an additional function of integumentary structures in extinct archosaurs?. *Historical Biology*, 33:1718–1725.

59 Zelenitsky, D.K., Therrien, F., Ridgely, R.C., McGee, R.A., and Witmer, L.M. 2011. Evolution of olfaction in non-avian theropod dinosaurs and birds. *Proceedings of the Royal Society B: Biological Sciences*, 278:3625–3634.

60 Sakagami, R. and Kawabe, S. 2020. Endocranial anatomy of the ceratopsid dinosaur *Triceratops* and interpretations of sensory and motor function. *PeerJ*, 8:e9888.

61 Stevens, K.A. 2006. Binocular vision in theropod dinosaurs. *Journal of Vertebrate Paleontology*, 26:321–330.

62 Choiniere, J.N., Neenan, J.M., Schmitz, L., Ford, D.P., Chapelle, K.E., Balanoff, A.M., Sipla, J.S., Georgi, J.A., Walsh, S.A., Norell, M.A., and Xu, X. 2021. Evolution of vision and hearing modalities in theropod dinosaurs. *Science*, 372:610–613.

63 Barker, C.T., Naish, D., Newham, E., Katsamenis, O.L,. and Dyke, G. 2017. Complex neuroanatomy in the rostrum of the Isle of Wight theropod *Neovenator salerii*. *Scientific Reports*, 7:1–8.

64 Bronzati, M., Rauhut, O.W., Bittencourt, J.S. and Langer, M.C. 2017. Endocast of the Late Triassic (Carnian) dinosaur *Saturnalia tupiniquim*: implications for the evolution of brain tissue in Sauropodomorpha. *Scientific Reports*, 7:1–7.

65 Paulina-Carabajal, A., and Currie, P.J. 2017. The braincase of the theropod dinosaur *Murusraptor*: osteology, neuroanatomy and comments on the paleobiological implications of certain endocranial features. *Ameghiniana*, 54(5):617–640.

66 Paulina-Carabajal, A., Bronzati, M., and Cruzado-Caballero, P. 2022. Paleoneurology of non-avian dinosaurs: an overview. In: *Paleoneurology of Amniotes: New Directions in the Study of Fossil Endocasts*, 267–332. Springer, Cham.

67 McKeown, M., Brusatte, S.L., Williamson, T.E., Schwab, J.A., Carr, T.D., Butler, I.B., Muir, A., Schroeder, K., Espy, M.A., Hunter, J.F., and Losko, A.S. 2020. Neurosensory and sinus evolution as tyrannosauroid dinosaurs developed giant size: insight from the endocranial anatomy of *Bistahieversor sealeyi*. *The Anatomical Record*, 303:1043–1059.

68 Hurlburt, G.R. 1996. Relative brain size in recent and fossil amniotes: determination and interpretation, Ph.D. thesis, University of Toronto.

69 Evans, D.C., Ridgely, R., and Witmer, L.M. 2009. Endocranial anatomy of lambeosaurine hadrosaurids (Dinosauria: Ornithischia): A sensorineural perspective on cranial crest function. *The Anatomical Record*, 292:1315–1337.

70 Evans, D.C. 2005. New evidence on brain–endocranial cavity relationships in ornithischian dinosaurs. *Acta Palaeontologica Polonica*, 50:617–622.

71 Müller, R.T., Ferreira, J.D., Pretto, F.A., Bronzati, M., and Kerber, L. 2021. The endocranial anatomy of *Buriolestes schultzi* (Dinosauria: Saurischia) and the early evolution of brain tissues in sauropodomorph dinosaurs. *Journal of Anatomy*, 238:809–827.

72 Jirak, D., and Janacek, J. 2017. Volume of the crocodilian brain and endocast during ontogeny. *PLoS One*, 12:e0178491.

73 Knoll, F., Lautenschlager, S., Kawabe, S., Martínez, G., Espílez, E., Mampel, L., and Alcalá, L. 2021. Palaeoneurology of the early cretaceous iguanodont *Proa valdearinnoensis* and its bearing on the parallel developments of cognitive abilities in theropod and ornithopod dinosaurs. *Journal of Comparative Neurology*, 529:3922–3945.

74 van Schaik, C.P., Triki, Z., Bshary, R., and Heldstab, S.A. 2021. A farewell to the encephalization quotient: a new brain size measure for comparative primate cognition. *Brain, Behavior and Evolution*, 96:1–12.

75 Healy, S.D. 2021. *Adaptation and the Brain*. Oxford University Press, Oxford.

76 Healy, S.D., and Rowe, C. 2007. A critique of comparative studies of brain size. *Proceedings of the Royal Society B: Biological Sciences*, 274:453–464.

77 Logan, C.J., Avin, S., Boogert, N., Buskell, A., Cross, F.R., Currie, A., Jelbert, S., Lukas, D., Mares, R., Navarrete, A.F., and Shigeno, S. 2018. Beyond brain size: Uncovering the neural correlates of behavioral and cognitive specialization. *Comparative Cognition & Behavior Reviews*, 13:55–89.

78 Varricchio, D.J., Hogan, J.D., and Freimuth, W.J. 2021. Revisiting Russell's troodontid: autecology, physiology, and speculative tool use. *Canadian Journal of Earth Sciences*, 58:796–811.

79 Fastovsky, D.F., and Weishampel, D.B. 2005. *The Evolution and Extinction of the Dinosaurs*. Cambridge University Press, Cambridge.

80 Chiappe, L.M., Coria, R.A., Dingus, L., Jackson, F., Chinsamy, A., and Fox, M. 1998. Sauropod dinosaur embryos from the Late Cretaceous of Patagonia. *Nature*, 396:258–261.

81 Niedźwiedzki, G., Singer, T., Gierliński, G.D., and Lockley, M.G. 2012. A protoceratopsid skeleton with an associated track from the Upper Cretaceous of Mongolia. *Cretaceous Research*, 33:7–10.

82 Lockley, M.G. 1991. *Tracking Dinosaurs: A New Look at an Ancient World*. CUP Archive.

83 Behrensmeyer, A.K., Kidwell, S.M., and Gastaldo, R.A. 2000. Taphonomy and paleobiology. *Paleobiology*, 26:103–147.

84 Fiorillo, A.R., and Eberth, D.A. 2004. Dinosaur taphonomy. In: *The Dinosauria*, 2nd ed., 607–613. University of California Press, Berkeley.

85 Mallon, J.C., Henderson, D.M., McDonough, C.M., and Loughry, W.J. 2018. A "bloat-and-float" taphonomic model best explains the upside-down preservation of ankylosaurs. *Palaeogeography, Palaeoclimatology, Palaeoecology*, 497:117–127.

86 Cuff, A.R., and Rayfield, E.J. 2015. Retrodeformation and muscular reconstruction of ornithomimosaurian dinosaur crania. *PeerJ*, 3:e1093.

87 Benton, M.J., Dunhill, A.M., Lloyd, G.T., and Marx, F.G. 2011. Assessing the quality of the fossil record: insights from vertebrates. *Geological Society, London, Special Publications*, 358:63–94.

88 Hone, D.W., and Rauhut, O.W. 2010. Feeding behaviour and bone utilization by theropod dinosaurs. *Lethaia*, 43:232–244.

89 Brown, C.M., Campione, N.E., Mantilla, G.P.W., and Evans, D.C. 2021. Size-driven preservational and macroecological biases in the latest Maastrichtian terrestrial vertebrate assemblages of North America. *Paleobiology*, 48:210–238.

90 Britt, B.B., Eberth, D.A., Scheetz, R.D., Greenhalgh, B.W., and Stadtman, K.L. 2009. Taphonomy of debris-flow hosted dinosaur bonebeds at Dalton Wells,

Utah (Lower Cretaceous, Cedar Mountain Formation, USA). *Palaeogeography, Palaeoclimatology, Palaeoecology*, 280:1–22.

91 Ryan, M.J., Russell, A.P., Eberth, D.A., and Currie, P.J. 2001. The taphonomy of a *Centrosaurus* (Ornithischia: Cer[a]topsidae) bone bed from the Dinosaur Park Formation (Upper Campanian), Alberta, Canada, with comments on cranial ontogeny. *Palaios*, 16:482–506.

92 Caro, T.M. 1994. *Cheetahs of the Serengeti Plains: Group Living in an Asocial Species.* Chicago: University of Chicago Press.

93 Tinbergen, N. 1963. On aims and methods in ethology. *Zeitschrift für Tierpsychologie*, 20:410–433.

94 Barnard, C.J. 2012. *Animal Behaviour: Ecology and Evolution.* Springer Science & Business Media.

95 Funston, P.J., Mills, M.G.L., Biggs, H.C., and Richardson, P.R.K. 1998. Hunting by male lions: ecological influences and socioecological implications. *Animal Behaviour*, 56:1333–1345.

96 Pauwels, O.S., Barr, B., Sanchez, M.L., and Burger, M. 2007. Diet records for the dwarf crocodile (*Osteolaemus tetraspis tetraspis*) in Rabi Oil Fields and Loango National Park, Southwestern Gabon. *Hamadryad*, 31:258–264.

97 Hone, D.W.E., and Faulkes, C.G. 2014. A proposed framework for establishing and evaluating hypotheses about the behaviour of extinct organisms. *Journal of Zoology*, 292:260–267.

98 Costa, J.T., and T.D. Fitzgerald. 1996. Social terminology revisited: Where are we ten years later? *Annales Zoologici Fennici*, 42:559–564.

99 Naish, D. 2014. The fossil record of bird behaviour. *Journal of Zoology*, 292:268–280.

100 Helm, B., Piersma, T., and Van der Jeugd, H. 2006. Sociable schedules: interplay between avian seasonal and social behaviour. *Animal Behaviour*, 72:245–262.

101 Shoshani, J., and J.F. Eisenberg. 1982. *Elephas maximus. Mammalian species*, 182:1–8.

102 Varricchio, D.J. 1995. Taphonomy of Jack's Birthday Site, a diverse dinosaur bonebed from the Upper Cretaceous Two Medicine Formation of Montana. *Palaeogeography, Palaeoclimatology, Palaeoecology*, 114:297–323.

103 Witmer, L.M. 1995. The extant phylogenetic bracket and the importance of reconstructing soft tissues in fossils. *Functional Morphology in Vertebrate Paleontology*, 1:19–33.

104 Brazaitis, P., and Watanabe, M.E. 2011. Crocodilian behaviour: a window to dinosaur behaviour?. *Historical Biology*, 23: 73–90.

105 Doody, J.S., Dinets, V., and Burghardt, G.M. 2021. *The Secret Social Lives of Reptiles.* Johns Hopkins University Press, Baltimore.

106 Snively, E., and Russell, A.P. 2003. Kinematic model of tyrannosaurid (Dinosauria: Theropoda) arctometatarsus function. *Journal of Morphology*, 255:215–227.

107 Xu, X., Wang, D.Y., Sullivan, C., Hone, D.W., Han, F.L., Yan, R.H., and Du, F.M. 2010. A basal parvicursorine (Theropoda: Alvarezsauridae) from the Upper Cretaceous of China. *Zootaxa*, 2413:1–19.

108 Janis, C. 1982. Evolution of horns in ungulates: ecology and paleoecology. *Biological Reviews*, 57: 261–318.

109 Carrano, M.T., C.M. Janis, and J.J. Sepkoski. 1999. Hadrosaurs as ungulate parallels: lost lifestyles and deficient data. *Acta Palaeontologica Polonica*, 44:237–261.

110 Godefroit, P., Sinitsa, S.M., Dhouailly, D., Bolotsky, Y.L., Sizov, A.V., McNamara, M.E., Benton, M.J., and Spagna, P. 2014. A Jurassic ornithischian dinosaur from Siberia with both feathers and scales. *Science*, 345:451–455.

111 Hone, D.W.E., and Mallon, J.C. 2017. Protracted growth impedes the detection of sexual dimorphism in non-avian dinosaurs. *Palaeontology*, 60:535–545.

112 Brusatte, S.L., Norell, M.A., Carr, T.D., Erickson, G.M., Hutchinson, J.R., Balanoff, A.M., Bever, G.S., Choiniere, J.N., Makovicky, P.J., and Xu, X. 2010. Tyrannosaur paleobiology: new research on ancient exemplar organisms. *Science*, 329:1481–1485.

113 DePalma, R.A., Burnham, D.A., Martin, L.D., Rothschild, B.M., and Larson, P.L. 2013. Physical evidence of predatory behavior in *Tyrannosaurus rex*. *Proceedings of the National Academy of Sciences*, 110(31):12560–12564.

114 Hone, D.W., and Watabe, M. 2010. New information on scavenging and selective feeding behaviour of tyrannosaurids. *Acta Palaeontologica Polonica*, 55:627–634.

115 Erickson, G.M. 2014. On dinosaur growth. *Annual Review of Earth and Planetary Sciences*, 42:675–697.

116 Henderson, D.M., and Snively, E. 2004. *Tyrannosaurus* en pointe: allometry minimized rotational inertia of large carnivorous dinosaurs. *Proceedings of the Royal Society of London B: Biological Sciences*, 271:S57–S60.

117 Hone, D. 2016. *The Tyrannosaur Chronicles: The Biology of the Tyrant Dinosaurs*. Bloomsbury Publishing.

118 Larson, P.L. 2008. Variation and sexual dimorphism in *Tyrannosaurus rex*. In: Tyrannosaurus rex: *The Tyrant King*, 103–128. Indiana University Press, Bloomington.

119 Mallon, J.C. 2017. Recognizing sexual dimorphism in the fossil record: lessons from nonavian dinosaurs. *Paleobiology*, 43:495–507.

120 Currie, P.J., and D.A. Eberth. 2010. On gregarious behavior in *Albertosaurus*. *Canadian Journal of Earth Sciences*, 47:1277–1289.

121 Tanke, D.H., and Currie, P.J. 1998. Head–biting behavior in theropod dinosaurs: paleopathological evidence. *Gaia*, 15:167–184.

122 Erickson, G.M., Currie, P.J., Inouye, B.D., and Winn, A.A. 2006. Tyrannosaur life tables: an example of nonavian dinosaur population biology. *Science*, 313:213–217.

123 Padian, K. 2022. Why tyrannosaurid forelimbs were so short: An integrative hypothesis. *Acta Palaeontologica Polonica*, 67:63–76.

124 Krauss, D., and Robinson, J. 2013. The biomechanics of a plausible hunting strategy for *Tyrannosaurus rex*. In: *Tyrannosaurid Paleobiology*, 251–264. Indiana University Press, Bloomington.

125 Lipkin, C. and Carpenter, K., 2008. Looking again at the forelimb of *Tyrannosaurus rex*. In: Tyrannosaurus rex: *The Tyrant King*, 166–190. Indiana University Press, Bloomington.

126 Hall, M.I., Kamilar, J.M., and Kirk, E.C. 2012. Eye shape and the nocturnal bottleneck of mammals. *Proceedings of the Royal Society B: Biological Sciences*, 279:4962–4968.

127 Schmitz, L., and Motani, R. 2011. Nocturnality in dinosaurs inferred from scleral ring and orbit morphology. *Science*, 332:705–708.

128 Hartman, S., Lovelacs, D., Linzmeier, B., Mathewson, P., and Porter, W. 2022. Mechanistic thermal modeling of Late Triassic terrestrial amniotes predicts biogeographic distribution. *Diversity*, 13:973.

129 Walsh, S.A., Barrett, P.M., Milner, A.C., Manley, G. and Witmer, L.M. 2009. Inner ear anatomy is a proxy for deducing auditory capability and behaviour in reptiles and birds. *Proceedings of the Royal Society of London B: Biological Sciences*, 276:1355–1360.

130 Carr, T.D., Varricchio, D.J. Sedlmayr, J.C., Roberts, E.M. and Moore, J.R. 2017. A new tyrannosaur with evidence for anagenesis and crocodile-like facial sensory system. *Scientific Reports*, 7:44942.

131 Bell, P.R., and Hendrickx, C. 2021. Epidermal complexity in the theropod dinosaur *Juravenator* from the Upper Jurassic of Germany. *Palaeontology*, 64:203–223.

132 Owen-Smith, N. 2021. Niche distinctions: resources versus risks. In: *Only in Africa: The Ecology of Human Evolution*, 145–169. Cambridge University Press.

133 Farlow, J.O., Dodson, P., and Chinsamy, A. 1995. Dinosaur biology. *Annual Review of Ecology and Systematics*, 26:445–471.

134 Arbour, V.M., Zanno, L.E., and Gates, T. 2016. Ankylosaurian dinosaur palaeoenvironmental associations were influenced by extirpation, sea-level fluctuation, and geodispersal. *Palaeogeography, Palaeoclimatology, Palaeoecology*, 449:289–299.

135 Rogers, R.R., Regan, A.K., Weaver, L.N., Thole, J.T., and Fricke, H.C. 2020. Tracking authigenic mineral cements in fossil bones from the Upper Cretaceous (Campanian) Two Medicine and Judith River formations, Montana. *Palaios*, 35:135–150.

136 Brinkman, D.B., Ryan, M.J., and Eberth, D.A. 1998. Paleogeographic and stratigraphic distribution of Ceratopsids in the Upper Judith River Group of Western Canada. *Palaios*, 13:160–169.

137 Tütken, T., Sander, M., Hummel, J., and Gee, C. 2007. Ernährung und Mobilität von Sauropoden—Informationspotential der Isotopenzusammensetzung von Knochen und Zähnen. *Hallesches Jahrbuch der Geowissenschaften*, Beiheft, 23:85–92.

138 Fricke, H.C., and Pearson, D.A. 2008. Stable isotope evidence for changes in dietary niche partitioning among hadrosaurian and ceratopsian dinosaurs of the Hell Creek Formation, North Dakota. *Paleobiology*, 34:534–552.

139 Chinsamy, A., Thomas, D.B., Tumarkin-Deratzian, A.R., and Fiorillo, A.R. 2012. Hadrosaurs were perennial polar residents. *The Anatomical Record*, 295:610–614.

140 Leuthold, W. 1977. The influence of environmental factors on the spatial and social organization. In: *African Ungulates: A Comparative Review of Their Ethology and Behavioral Ecology*. Springer, Berlin.

141 Owen-Smith, N., Hopcraft, G., Morrison, T., Chamaillé-Jammes, S., Hetem, R., Bennitt, E., and Van Langevelde, F. 2020. Movement ecology of large herbivores in African savannas: current knowledge and gaps. *Mammal Review*, 50(3):252–266.

142 Endler, J.A. 1992. Signals, signal conditions, and the direction of evolution. *The American Naturalist*, 139:S125–S153.

143 Grady, J.M., Enquist, B.J., Dettweiler-Robinson, E., Wright, N.A., and Smith, F.A. 2014. Evidence for mesothermy in dinosaurs. *Science*, 344:1268–1272.

144 Lovelace, D.M., Hartman, S.A., Mathewson, P.D., Linzmeier, B.J., and Porter, W.P. 2020. Modeling Dragons: Using linked mechanistic physiological and microclimate models to explore environmental, physiological, and morphological constraints on the early evolution of dinosaurs. *PLoS One*, 15:e0223872.

145 Olsen, P., Sha, J., Fang, Y., Chang, C., Whiteside, J.H., Kinney, S., Sues, H.D., Kent, D., Schaller, M., and Vajda, V. 2022. Arctic ice and the ecological rise of the dinosaurs. *Science Advances*, 8:eabo6342.

146 Taylor, M.P., Wedel, M.J., and Naish, D. 2009. Head and neck posture in sauropod dinosaurs inferred from extant animals. *Acta Palaeontologica Polonica*, 54:213–220.

147 Mallison, H. 2010. The digital *Plateosaurus* II: an assessment of the range of motion of the limbs and vertebral column and of previous reconstructions using a digital skeletal mount. *Acta Palaeontologica Polonica*, 55:433–458.

148 Mallison, H. 2011. *Plateosaurus* in 3D: how CAD models and kinetic–dynamic modeling bring an extinct animal to life. In: *Biology of the Sauropod Dinosaurs: Understanding the Life of Giants*, 219–236. Indiana University Press, Bloomington.

149 Senter, P. 2006. Comparison of forelimb function between *Deinonychus* and *Bambiraptor* (Theropoda: Dromaeosauridae). *Journal of Vertebrate Paleontology*, 26:897–906.

150 Milner, A.R., Harris, J.D., Lockley, M.G., Kirkland, J.I., and Matthews, N.A. 2009. Bird-like anatomy, posture, and behavior revealed by an Early Jurassic theropod dinosaur resting trace. *PloS One*, 4:e4591.

151 Xu, X.. and Norell, M.A. 2004. A new troodontid dinosaur from China with avian-like sleeping posture. *Nature*, 431:838–841.

152 Rogers, C.S., Hone, D.W., McNamara, M.E., Zhao, Q., Orr, P.J., Kearns, S.L., and Benton, M.J. 2015. The Chinese Pompeii? Death and destruction of dinosaurs in the Early Cretaceous of Lujiatun, NE China. *Palaeogeography, Palaeoclimatology, Palaeoecology*, 427:89–99.

153 Bishop, P.J., Cuff, A.R., and Hutchinson, J.R. 2021. How to build a dinosaur: Musculoskeletal modeling and simulation of locomotor biomechanics in extinct animals. *Paleobiology*, 47:1–38.

154 Hutchinson, J.R., and Garcia, M. 2002. *Tyrannosaurus* was not a fast runner. *Nature*, 415:1018–1021.

155 Dececchi, T.A., Larsson, H.C., and Habib, M.B. 2016. The wings before the bird: an evaluation of flapping-based locomotory hypotheses in bird antecedents. *PeerJ*, 4:p.e2159.

156 Senter, P. 2005. Function in the stunted forelimbs of *Mononykus olecranus* (Theropoda), a dinosaurian anteater. *Paleobiology*, 31:373–381.

157 Fowler, D.W., and Hall, L.E. 2011. Scratch-digging sauropods, revisited. *Historical Biology*, 23:27–40.

158 Woodruff, D.C., and Varricchio, D.J. 2011. Experimental modeling of a possible *Oryctodromeus cubicularis* (Dinosauria) burrow. *Palaios*, 26:140–151.

159 Cau, A., Beyrand, V., Voeten, D.F., Fernandez, V., Tafforeau, P., Stein, K., Barsbold, R., Tsogtbaatar, K., Currie, P.J., and Godefroit, P. 2017. Synchrotron scanning reveals amphibious ecomorphology in a new clade of bird—like dinosaurs. *Nature*, 552:395–399.

160 Hone, D.W.E., and Holtz, T.R. 2017. A century of spinosaurs - a review and revision of the Spinosauridae with comments on their ecology. *Acta Geologica Sinica* (English edition), 91:1120–1132.

161 Ibrahim, N., Sereno, P.C., Dal Sasso, C., Maganuco, S., Fabbri, M., Martill, D.M., Zouhri, S., Myhrvold, N., and Iurino, D.A. 2014. Semiaquatic adaptations in a giant predatory dinosaur. *Science*, 345:1613–1616.

162 Ibrahim, N., Maganuco, S., Dal Sasso, C., Fabbri, M., Auditore, M., Bindellini, G., Martill, D.M., Zouhri, S., Mattarelli, D.A., Unwin, D.M., and Wiemann, J. 2020. Tail-propelled aquatic locomotion in a theropod dinosaur. *Nature*, 581:67–70.

163 Fabbri, M., Navalón, G., Benson, R.B., Pol, D., O'Connor, J., Bhullar, B.A.S., Erickson, G.M., Norell, M.A., Orkney, A., Lamanna, M.C., and Zouhri, S. 2022. Subaqueous foraging among carnivorous dinosaurs. *Nature*, 603:852–857.

164 Hone, D.W., and Holtz, T.R. 2021. Evaluating the ecology of *Spinosaurus*: Shoreline generalist or aquatic pursuit specialist?. *Palaeontologia Electronica*, 24:a03.

165 Sereno, P.C., Myhrvold, N., Henderson, D.M., Fish, F.E., Vidal, D., Baumgart, S.L., Keillor, T.M., Formoso, K.K., and Conroy, L.L. 2022. *Spinosaurus* is not an aquatic dinosaur. *eLife*, 11:e80092.

166 Myhrvold, N.P., Baumgart, S.L., Vidal, D., Fish, F.E., Henderson, D.M., Saitta, E.T., and Sereno, P.C., 2024. Diving dinosaurs? Caveats on the use of bone compactness and pFDA for inferring lifestyle. *PLoS ONE*, 19:p.e0298957.

167 Barker, C.T., Naish, D., Trend, J., Michels, L.V., Witmer, L., Ridgley, R., Rankin, K., Clarkin, C.E., Schneider, P., and Gostling, N.J. 2023. Modified skulls but conservative brains? The palaeoneurology and endocranial anatomy of baryonychine dinosaurs (Theropoda: Spinosauridae). *Journal of Anatomy*, 242:1124–1145.

168 Amiot, R., Buffetaut, E., Lecuyer, C., Wang, X., Boudad, L., Ding, Z., Fourel, F., Hutt, S., Martineau, F., Medeiros, M.A., Mo, J., Simon, L., Suteethorn, V., Sweetman, S., Tong, H., Zhang, F., and Zhou, Z. 2010. Oxygen isotope evidence for semi-aquatic habits among spinosaurid theropods. *Geology*, 38:139–142.

169 Ostrom, J.H. 1972. Were some dinosaurs gregarious?. *Palaeogeography, Palaeo-climatology, Palaeoecology*, 11: 287–301.

170 Ostrom, J.H. 1990. Dromaeosauridae. In: D.B. Weishampel, P. Dodson and H. Osmólska (eds.), *The Dinosauria*, 269–279. University of California Press, Berkeley.

171 Taborsky, M., Cant, M.A., and Komdeur, J. 2021. *The Evolution of Social Behaviour*. Cambridge University Press.

172 Sampson, S.D. 2001. Speculations on the socioecology of Ceratopsid dinosaurs (Ornithischia: Neoceratopsia). In: *Mesozoic Vertebrate Life*, 263–276. Indiana University Press, Bloomington.

173 De Ruiter, J.R. 1986. The influence of group size on predator scanning and for-aging behaviour of wedgecapped capuchin monkeys (*Cebus olivaceus*). *Behaviour*, 98:240–258.

174 Hill, R.A., and P.C. Lee. 1998. Predation risk as an influence on group size in cercopithecoid primates: implications for social structure. *Journal of Zoology*, 245:447–456.

175 Burton, R. 1985. *Bird Behaviour*. Granada Publishing. London

176 Krause, J., and Ruxton, G.D. 2002. *Living in Groups*. Oxford University Press, Oxford.

177 Mosser, A., and C. Packer. 2009. Group territoriality and the benefits of sociality in the African lion, *Panthera leo*. *Animal Behaviour*, 78:359–370.

178 Shen, S.F., Emlen, S.T., Koenig, W.D., and Rubenstein, D.R. 2017. The ecology of cooperative breeding behaviour. *Ecology Letters*, 20:708–720.

179 Koenig, W.D. 1981. Reproductive success, group size, and the evolution of coop-erative breeding in the acorn woodpecker. *The American Naturalist*, 117:421–443.

180 Majolo, B., de Bortoli Vizioli, A., and Schino, G. 2008. Costs and benefits of group living in primates: group size effects on behaviour and demography. *Animal Behaviour*, 76:1235–1247.

181 Burghardt, G.M. 1977. Of iguanas and dinosaurs: Social behavior and commu-nication in neonate reptiles. *American Zoologist*, 17:177–190.

182 Coetzee, H.C. 2010. Observations of southern ground-hornbill *Bucorvus leadbeateri* grooming common warthog *Phacochoerus africanus*. *African Journal of Ecology*, 48:1131–1133.

183 Stensland, E., Angerbjörn, A., and Berggren, P. 2003. Mixed species groups in mammals. *Mammal Review*, 33:205–223.

184 Dinets, V., and Eligulashvili, B. 2016. Striped Hyaenas (*Hyaena hyaena*) in Grey Wolf (*Canis lupus*) packs: cooperation, commensalism or singular aberration?. *Zoology in the Middle East*, 62:85–87.

185 Ashton, B.J., Ridley, A.R., Edwards, E.K., and Thornton, A. 2018. Cognitive performance is linked to group size and affects fitness in Australian magpies. *Nature*, 554:364–367.

186 Shultz, S., and Dunbar, R.I.M. 2006. Both social and ecological factors predict ungulate brain size. *Proceedings of the Royal Society B: Biological Sciences*, 273:207–215.

187 Shultz, S., and Dunbar, R.I.M. 2010. Social bonds in birds are associated with brain size and contingent on the correlated evolution of life-history and increased parental investment. *Biological Journal of the Linnean Society*, 100:111–123.

188 Lang, J.W. 1987. Crocodilian behaviour: implications for management. In: *Wildlife Management: Crocodiles and Alligators*, 273–294. Surrey Beatty, Sydney.

189 Chiba, K., Ryan, M.J., Braman, D.R., Eberth, D.A., Scott, E.E., Brown, C.M., Kobayashi,Y., and Evans, D.C. 2015. Taphonomy of a monodominant *Centrosaurus apertus* (Dinosauria: Ceratopsia) bonebed from the upper Oldman Formation of southeastern Alberta. *Palaios*, 30:655–667.

190 Heredia, A.M., Díaz-Martínez, I., Pazos, P.J., Comerio, M., and Fernández, D.E. 2020. Gregarious behaviour among non-avian theropods inferred from trackways: A case study from the Cretaceous (Cenomanian) Candeleros Formation of Patagonia, Argentina. *Palaeogeography, Palaeoclimatology, Palaeoecology*, 538:109480.

191 Voorhies, M.R. 1985. A Miocene rhinoceros herd buried in volcanic ash. *National Geographic Society Research Reports*, 19:671–688.

192 Martill, D.M. and Naish, D. 2001. *Dinosaurs of the Isle of Wight*. Palaeontological Association, London.

193 Getty, P.R., Aucoin, C., Fox, N., Judge, A., Hardy, L., and Bush, A.M. 2017. Perennial lakes as an environmental control on theropod movement in the Jurassic of the Hartford Basin. *Geosciences*, 7:13.

194 Richter, A., and Böhme, A. 2016. Too many tracks: preliminary description and interpretation of the diverse and heavily dinoturbated Early Cretaceous "Chicken Yard" ichnoassemblage (Obernkirchen Tracksite, northern Germany). In: *Dinosaur Tracks: The Next Steps*, 334–357. Indiana University Press, Bloomington.

195 Rinehart, L.F., Lucas, S.G., Heckert, A.B., Spielmann, J.A., and Celeskey, M.B. 2009. The Paleobiology of *Coelophysis bauri* (Cope) from the Upper Triassic (Apachean) Whitaker quarry, New Mexico, with detailed analysis of a single quarry block. *New Mexico Museum of Natural History Science Bulletin*, 45:1–260.

196 Cotton, W.D., Cotton, J.E., and Hunt, A.P. 1998. Evidence for social behavior in ornithopod dinosaurs from the Dakota Group of northeastern New Mexico, USA. *Ichnos*, 6:141–149.

197 Persons IV, W.S., Funston, G.F., Currie, P.J., and Norell, M.A. 2015. A possible instance of sexual dimorphism in the tails of two oviraptorosaur dinosaurs. *Scientific Reports*, 5:9472.

198 Botfalvai, G., Prondvai, E., and Ősi, A. 2020. Living alone or moving in herds? A holistic approach highlights complexity in the social lifestyle of Cretaceous ankylosaurs. *Cretaceous Research*, 118:104633.

199 Varricchio, D.J., Sereno, P.C., Xijin, Z., Lin, T., Wilson, J.A., and Lyon, G.H. 2008. Mud-trapped herd captures evidence of distinctive dinosaur sociality. *Acta Palaeontologica Polonica*, 53:567–578.

200 Myers, T.S., and Fiorillo, A.R. 2009. Evidence for gregarious behavior and age segregation in sauropod dinosaurs. *Palaeogeography, Palaeoclimatology, Palaeoecology*, 274:96–104.

201 Pol, D., Mancuso, A.C., Smith, R.M., Marsicano, C.A., Ramezani, J., Cerda, I.A., Otero, A., and Fernandez, V. 2021. Earliest evidence of herd-living and age segregation amongst dinosaurs. *Scientific Reports*, 11(1):1–9.

202 Funston, G.F., Currie, P.J., Eberth, D.A., Ryan, M.J., Chinzorig, T., Badamgarav, D., and Longrich, N.R. 2016. The first oviraptorosaur (Dinosauria: Theropoda) bonebed: evidence of gregarious behaviour in a maniraptoran theropod. *Scientific Reports*, 6:1–13.

203 Lockley, M., Mitchell, L., and Odier, G.P. 2007. Small theropod track assemblages from Middle Jurassic eolianites of Eastern Utah: paleoecological insights from dune ichnofacies in a transgressive sequence. *Ichnos*, 14:131–142.

204 Kirkland, J.I., Simpson, E.L., DeBlieux, D.D., Madsen, S.K., Bogner, E., and Tibert, N.E. 2016. Depositional constraints on the Lower Cretaceous Stikes Quarry dinosaur site: Upper Yellow Cat Member, Cedar Mountain Formation, Utah. *Palaios*, 31:421–439.

205 Sander, P.M. 1992. The Norian *Plateosaurus* bonebeds of central Europe and their taphonomy. *Palaeogeography, Palaeoclimatology, Palaeoecology*, 93:255–299.

206 Coria, R.A. 1994. On a monospecific assemblage of sauropod dinosaurs from Patagonia: implications for gregarious behavior. *GAIA: Revista de Geociências*, 10:209–213.

207 Storrs, G.W., Oser, S.E., and Aull, M. 2012. Further analysis of a Late Jurassic dinosaur bone-bed from the Morrison Formation of Montana, USA, with a computed three-dimensional reconstruction. *Earth and Environmental Science Transactions of the Royal Society of Edinburgh*, 103:443–458.

208 Day, J.J., Upchurch, P., Norman, D.B., Gale, A.S., and Powell, H.P. 2002. Sauropod trackways, evolution, and behavior. *Science*, 296:1659.

209 Wosik, M., and Evans, D.C. 2022. Osteohistological and taphonomic life-history assessment of *Edmontosaurus annectens* (Ornithischia: Hadrosauridae) from the Late Cretaceous (Maastrichtian) Ruth Mason dinosaur quarry, South Dakota, United States, with implication for ontogenetic segregation between juvenile and adult hadrosaurids. *Journal of Anatomy*, 241:272–296.

210 Matsukawa, M., Matsui, T., and Lockley, M.G. 2001. Trackway evidence of herd structure among ornithopod dinosaurs from the Cretaceous Dakota group of northeastern New Mexico, USA. *Ichnos*, 8:197–206.

211 Fiorillo, A.R., Hasiotis, S.T., and Kobayashi, Y. 2014. Herd structure in Late Cretaceous polar dinosaurs: A remarkable new dinosaur tracksite, Denali National Park, Alaska, USA. *Geology*, 42:719–722.

212 Hunt, R., and Farke, A. 2010. Behavioral interpretations from ceratopsid bonebeds. In: *New Perspectives on Horned Dinosaurs: The Royal Tyrrell Museum Ceratopsian Symposium*, 447–455. Indiana University Press, Bloomington.

213 Mathews, J.C., Brusatte, S.L., Williams, S.A., and Henderson, M.D. 2009. The first *Triceratops* bonebed and its implications for gregarious behavior. *Journal of Vertebrate Paleontology*, 29:286–290.

214 Hone, D.W., Farke, A.A., Watabe, M., Shigeru, S., and Tsogtbaatar, K. 2014. A new mass mortality of juvenile *Protoceratops* and size-segregated aggregation behaviour in juvenile non-avian dinosaurs. *PloS One,* 9:e113306.

215 Kobayashi, Y., and Lü, J.C. 2003. A new ornithomimid dinosaur with gregarious habits from the Late Cretaceous of China. *Acta Palaeontologica Polonica*, 48:235–259.

216 Myers, T.S. and Storrs, G.W. 2007. Taphonomy of the Mother's Day Quarry, Upper Jurassic Morrison Formation, south-central Montana, USA. *Palaios*, 22:651–666.

217 Forster, C.A. 1990. Evidence for juvenile groups in the ornithopod dinosaur *Tenontosaurus tilletti* Ostrom. *Journal of Paleontology*, 64:164–165.

218 Zhao, Q., Benton, M.J., Xu, X., and Sander, P.M. 2013. Juvenile-only clusters and behaviour of the Early Cretaceous dinosaur *Psittacosaurus*. *Acta Palaeontologica Polonica*, 59:827–833.

219 Arbour, V.M., Burns, M.E., Bell, P.R., and Currie, P.J. 2014. Epidermal and dermal integumentary structures of ankylosaurian dinosaurs. *Journal of Morphology*, 275:39–50.

220 Roach, B.T., and D.L. Brinkman. 2007. A reevaluation of cooperative pack hunting and gregariousness in *Deinonychus antirrhopus* and other nonavian theropod dinosaurs. *Bulletin of the Peabody Museum of Natural History*, 48:103–138.

221 Maxwell, W.D., and J.H. Ostrom. 1995. Taphonomy and paleobiological implications of *Tenontosaurus–Deinonychus* associations. *Journal of Vertebrate Paleontology*, 15:707–712.

222 Okarma, H., Jędrzejewski, W., Schmidt, K., Kowalczyk, R., and Jędrzejewska, B., 1997. Predation of Eurasian lynx on roe deer and red deer in Białowieża Primeval Forest, Poland. *Acta Theriologica*, 42(2):203–224.

223 Smith, J.M., and Harper, D. 2003. *Animal Signals*. Oxford University Press.

224 Galef, B.G., and Giraldeau, L.A. 2001. Social influences on foraging in vertebrates: causal mechanisms and adaptive functions. *Animal Behaviour*, 61:3–15.

225 Knell, R.J., and S.D. Sampson. 2011. Bizarre structures in dinosaurs. *Journal of Zoology*, 283:18–22.

226 Fitzgibbon, C.D., and Fanshawe, J.H. 1988. Stotting in Thomson's gazelles: an honest signal of condition. *Behavioral Ecology and Sociobiology*, 23:69–74.

227 Laidre, M.E., and Johnstone, R.A. 2013. Animal signals. *Current Biology*, 23:R829–R833.

228 Knell, R.J., Naish, D., Tomkins, J.L., and Hone, D.W. 2013. Sexual selection in prehistoric animals: detection and implications. *Trends in Ecology & Evolution*, 28:38–47.

229 McCullough, E.L., Miller, C.W., and Emlen, D.J. 2016. Why sexually selected weapons are not ornaments. *Trends in Ecology and Evolution*, 31:742–751.

230 Tobias, J.A., Montgomerie, R., and Lyon, B.E. 2012. The evolution of female ornaments and weaponry: social selection, sexual selection and ecological competition. *Philosophical Transactions of the Royal Society B: Biological Sciences*, 367:2274–2293.

231 Shuker, D.M., and Kvarnemo, C. 2021. The definition of sexual selection. *Behavioral Ecology*, 32:781–794.

232 Padian, K., and Horner, J.R. 2011. The evolution of 'bizarre structures' in dinosaurs: biomechanics, sexual selection, social selection or species recognition?. *Journal of Zoology*, 283: 3–17.

233 Hone, D.W.E., and Naish, D. 2013. The 'species recognition hypothesis' does not explain the presence and evolution of exaggerated structures in non-avialan dinosaurs. *Journal of Zoology*, 290:172–180.

234 Knapp, A., Knell, R.J., Farke, A.A., Loewen, M.A., and Hone, D.W. 2018. Patterns of divergence in the morphology of ceratopsian dinosaurs: sympatry is not a driver of ornament evolution. *Proceedings of the Royal Society B: Biological Sciences*, 285:20180312.

235 Saitta, E.T., Stockdale, M.T., Longrich, N.R., Bonhomme, V., Benton, M.J., Cuthill, I.C., and Makovicky, P.J. 2020. An effect size statistical framework for investigating sexual dimorphism in non-avian dinosaurs and other extinct taxa. *Biological Journal of the Linnean Society*, 131:231–273.

236 Erickson, G.M., Rogers, K.C., Varricchio, D.J., Norell, M.A., and Xu, X. 2007. Growth patterns in brooding dinosaurs reveals the timing of sexual maturity in non-avian dinosaurs and genesis of the avian condition. *Biology Letters*, 3:558–561.

237 Kraaijeveld, K., Gregurke, J., Hall, C., Komdeur, C., and Mulder, R.A. 2004. Mutual ornamentation, sexual selection, and social dominance in the black swan. *Behavioral Ecology*, 15:380–389.

238 Jacobs, G.H., and Rowe, M.P. 2004. Evolution of vertebrate colour vision. *Clinical and Experimental Optometry*, 87:206–216.

239 Isles, T.E. 2009. The socio-sexual behaviour of extant archosaurs: implications for understanding dinosaur behaviour. *Historical Biology*, 21:139–214.

240 Bonduriansky, R., and Day, T. 2003. The evolution of static allometry in sexually selected traits. *Evolution*, 57:2450–2458

241 Dodson, P. 1976. Quantitative aspects of relative growth and sexual dimorphism in *Protoceratops*. *Journal of Paleontology*, 1976:929–940.

242 Weishampel, D.B. 2022. Franz Baron Nopcsa: A short life of research in evolutionary paleobiology and Albanology. *The Anatomical Record*, 306:1969–1975.

243 Evans, D.C. 2010. Cranial anatomy and systematics of *Hypacrosaurus altispinus*, and a comparative analysis of skull growth in lambeosaurine hadrosaurids (Dinosauria: Ornithischia). *Zoological Journal of the Linnean Society*, 159:398–434.

244 Main, R.P., De Ricqlès, A., Horner, J.R., and Padian, K. 2005. The evolution and function of thyreophoran dinosaur scutes: implications for plate function in stegosaurs. *Paleobiology*, 31:291–314.

245 Brown, C.M. 2017. An exceptionally preserved armored dinosaur reveals the morphology and allometry of osteoderms and their horny epidermal coverings. *PeerJ*, 5:e4066.

246 Gates, T.A., Organ, C., and Zanno, L.E. 2016. Bony cranial ornamentation linked to rapid evolution of gigantic theropod dinosaurs. *Nature Communications*, 7:12931.

247 Bell, P.R., Campione, N.E., Persons, W.S., Currie, P.J., Larson, P.L., Tanke, D.H., and Bakker, R.T. 2017. Tyrannosauroid integument reveals conflicting patterns of gigantism and feather evolution. *Biology Letters*, 13:20170092.

248 Xu, X., Zheng, X., and You, H. 2010. Exceptional dinosaur fossils show ontogenetic development of early feathers. *Nature*, 464:1338.

249 Zhang, F., Zhou, Z., Xu, X., Wang, X., and Sullivan, C. 2008. A bizarre Jurassic maniraptoran from China with elongate ribbon-like feathers. *Nature*, 455:1105–1108.

250 Hu, D., Clarke, J.A., Eliason, C.M., Qiu, R., Li, Q., Shawkey, M.D., Zhao, C., D'Alba, L., Jiang, J., and Xu, X. 2018. A bony-crested Jurassic dinosaur with evidence of iridescent plumage highlights complexity in early paravian evolution. *Nature Communications*, 9:217.

251 Senter, P. 2007. Necks for sex: sexual selection as an explanation for sauropod dinosaur neck elongation. *Journal of Zoology*, 271:45–53.

252 Taylor, M.P., Hone, D.W., Wedel, M.J., and Naish, D. 2011. The long necks of sauropods did not evolve primarily through sexual selection. *Journal of Zoology*, 285:150–161.

253 Cerda, I.A., Novas, F.E., Carballido, J.L., and Salgado, L. 2022. Osteohistology of the hyperelongate hemispinous processes of *Amargasaurus cazaui* (Dinosauria: Sauropoda): Implications for soft tissue reconstruction and functional significance. *Journal of Anatomy*, 240:1005–1019.

254 Merilaita, S., Scott-Samuel, N.E., and Cuthill, I.C. 2017. How camouflage works. *Philosophical Transactions of the Royal Society B: Biological Sciences*, 372:20160341.

255 Vinther, J., Nicholls, R., Lautenschlager, S., Pittman, M., Kaye, T.G., Rayfield, E., Mayr, G., and Cuthill, I.C. 2016. 3D camouflage in an ornithischian dinosaur. *Current Biology*, 26:2456–2462.

256 Riede, T., Eliason, C.M., Miller, E.H., Goller, F., and Clarke, J.A. 2016. Coos, booms, and hoots: The evolution of closed-mouth vocal behavior in birds. *Evolution*, 70:1734–1746.

257 Capshaw, G., Willis, K.L., Han, D., and Bierman, H.S. 2021. Reptile sound production and perception. In: *Neuroendocrine Regulation of Animal Vocalization*, 101–118. Academic Press.

258 Weishampel, D.B. 1981. Acoustic analyses of potential vocalization in lambeosaurine dinosaurs (Reptilia: Ornithischia). *Paleobiology*, 7:252–261.

259 Bro-Jørgensen, J., and Dabelsteen, T. 2008. Knee-clicks and visual traits indicate fighting ability in eland antelopes: multiple messages and back-up signals. *BMC Biology*, 6:1–8.

260 Miyashita, T., Arbour, V.M., Witmer, L.M., and Currie, P.J. 2011. The internal cranial morphology of an armoured dinosaur *Euoplocephalus* corroborated by X-ray computed tomographic reconstruction. *Journal of Anatomy*, 219:661–675.

261 Yoshida, J., Kobayashi, Y., and Norell, M.A. 2023. An ankylosaur larynx provides insights for bird-like vocalization in non-avian dinosaurs. *Communications Biology*, 6:152.

262 Nassif, J.P., Witmer, L.M., and Ridgely, R. 2018. Preliminary osteological evidence for secondary loss of a tympanic membrane in multiple clades of ornithischian dinosaurs. *Society of Vertebrate Paleontology Abstracts*, Albuquerque.

263 Clarke, J.A., Chatterjee, S., Li, Z., Riede, T., Agnolin, T., Goller, F., Isasi, M.P., Martinioni, D.R., Mussel, F.J., and Novas, F.E. 2016. Fossil evidence of the avian vocal organ from the Mesozoic. *Nature*, 538: 502–505.

264 Myhrvold, N.P., and Currie, P.J. 1997. Supersonic sauropods? Tail dynamics in the diplodocids. *Paleobiology*, 23:393–409.

265 Conti, S., Tschopp, E., Mateus, O., Zanoni, A., Masarati, P., and Sala, G. 2022. Multibody analysis and soft tissue strength refute supersonic dinosaur tail. *Scientific Reports*, 12:1–9.

266 Ferguson, G.W. 1977. Display and communications in reptiles: an historical perspective. *American Zoologist*, 17:167–176.

267 Caro, S.P., and Balthazart, J. 2010. Pheromones in birds: myth or reality?. *Journal of Comparative Physiology A*, 196:751–766.

268 Vinther, J., Nicholls, B., and Kelly, D.A. 2021. A cloacal opening in a non-avian dinosaur. *Current Biology*, 31:R182–R183.

269 Fastovsky, D.E., Weishampel, D.B., Watabe, M., Barsbold, R., Tsogtbaatar, K.H., and Narmandakh, P. 2011. A nest of *Protoceratops andrewsi* (Dinosauria, Ornithischia). *Journal of Paleontology*, 85:1035–1041.

270 Maiorino, L., Farke, A.A., Kotsakis, T., and Piras, P. 2015. Males resemble females: re-evaluating sexual dimorphism in *Protoceratops andrewsi* (Neoceratopsia, Protoceratopsidae). *PLoS One*, 10:e0126464.

271 Hone, D.W., Wood, D., and Knell, R.J. 2016. Positive allometry for exaggerated structures in the ceratopsian dinosaur *Protoceratops andrewsi* supports sociosexual signaling. *Palaeontologia Electronica*, 19:1–13.

272 Knapp, A., Knell, R.J., and Hone, D.W.E. 2021. Three-dimensional geometric morphometric analysis of the skull of *Protoceratops andrewsi* supports a sociosexual signalling role for the ceratopsian frill. *Proceedings of the Royal Society B*, 288:20202938.

273 Tereschenko, V.S., and Singer, T. 2013. Structural features of neural spines of the caudal vertebrae of protoceratopoids (Ornithischia: Neoceratopsia). *Paleontological Journal*, 47:618–630.

274 Kokko, H., and Johnstone, R.A. 2002. Why is mutual mate choice not the norm? Operational sex ratios, sex roles and the evolution of sexually dimorphic and monomorphic signalling. *Philosophical Transactions of the Royal Society B: Biological Sciences*, 357:319–330.

275 Gross, M.R. 1996. Alternative reproductive strategies and tactics: diversity within sexes. *Trends in Ecology & Evolution*, 11:92–98.

276 Pianka, E.R. 1970. On *r*- and *K*-selection. *The American Naturalist*, 104:592–597.

277 Emlen, S.T., and Oring, L.W. 1977. Ecology, sexual selection, and the evolution of mating systems. *Science*, 197:215–223.

278 Mallison, H. 2011. Rearing giants: kinetic–dynamic modeling of sauropod bipedal and tripodal poses. In: *Biology of the Sauropod Dinosaurs: Understanding the Life of Giants*, 237–250. Indiana University Press, Bloomington.

279 Sanger, T.J., Gredler, M.L., and Cohn, M.J. 2015. Resurrecting embryos of the tuatara, *Sphenodon punctatus*, to resolve vertebrate phallus evolution. *Biology Letters*, 11:20150694.

280 Norell, M.A., Wiemann, J., Fabbri, M., Yu, C., Marsicano, C.A., Moore-Nall, A., Varricchio, D.J., Pol, D., and Zelenitsky, D.K. 2020. The first dinosaur egg was soft. *Nature*, 583(7816):406–410.

281 Horner, J.R. 2000. Dinosaur reproduction and parenting. *Annual Review of Earth and Planetary Sciences*, 28:19–45.

282 Stein, K., Prondvai, E., Huang, T., Baele, J.M., Sander, P.M., and Reisz, R. 2019. Structure and evolutionary implications of the earliest (Sinemurian, Early Jurassic) dinosaur eggs and eggshells. *Scientific Reports*, 9:1–9.

283 Deeming, D.C. 2006. Ultrastructural and functional morphology of eggshells supports the idea that dinosaur eggs were incubated buried in a substrate. *Palaeontology*, 49:171–185.

284 Horner, J.R., and Makela, R. 1979. Nest of juveniles provides evidence of family structure among dinosaurs. *Nature*, 282:296.

285 Chiappe, L.M., Jackson, F., Coria, R.A., and Dingus, L. 2005. Nesting titanosaurs from Auca Mahuevo and adjacent sites. In: *The Sauropods: Evolution and Paleobiology*, 285–302. University of California Press, Berkeley.

286 Pu, H., Zelenitsky, D.K., Lü, J., Currie, P.J., Carpenter, K., Xu, L., Koppelhus, E.B., Jia, S., Xiao, L., Chuang, H., and Li, T. 2017. Perinate and eggs of a giant caenagnathid dinosaur from the Late Cretaceous of central China. *Nature Communications*, 8:1–9.

287 Wiemann, J., Yang, T.R., and Norell, M.A. 2018. Dinosaur egg colour had a single evolutionary origin. *Nature*, 563:555–558.

288 Yang, T.R., Chen, Y.H., Wiemann, J., Spiering, B., and Sander, P.M. 2018. Fossil eggshell cuticle elucidates dinosaur nesting ecology. *PeerJ*, 6:e5144.

289 Varricchio, D.J., Jackson, F., Borkowski, J.J., and Horner, J.R. 1997. Nest and egg clutches of the dinosaur *Troodon formosus* and the evolution of avian reproductive traits. *Nature*, 385:247–250.

290 Lack, D. 1947. The significance of clutch-size. *Ibis*, 89:302–352.

291 Bertram, B.C.R. 2014. *The Ostrich Communal Nesting System*. Vol. 11 in Monographs in Behavior and Ecology. Princeton University Press.

292 Martin, T.E. 2015. Age-related mortality explains life history strategies of tropical and temperate songbirds. *Science*, 349:966–970.

293 Böhning-Gaese, K., Halbe, B., Lemoine, N., and Oberrath, R. 2000. Factors influencing the clutch size, number of broods and annual fecundity of North American and European land birds. *Evolutionary Ecology Research*, 2:823–839.

294 Birchard, G.F., Ruta, M., and Deeming, D.C. 2013. Evolution of parental incubation behaviour in dinosaurs cannot be inferred from clutch mass in birds. *Biology Letters*, 9:20130036.

295 Bi, S., Amiot, R., de Fabrègues, C.P., Pittman, M., Lamanna, M.C., Yu, Y., Yu, C., Yang, T., Zhang, S., Zhao, Q., and Xu, X. 2021. An oviraptorid preserved atop an embryo-bearing egg clutch sheds light on the reproductive biology of non-avialan theropod dinosaurs. *Science Bulletin*, 66:947–954.

296 Hechenleitner, E.M., Grellet-Tinner, G., and Fiorelli, L.E. 2015. What do giant titanosaur dinosaurs and modern Australasian megapodes have in common?. *PeerJ*, 3:e1341.

297 Wesolowski, T. 1994. On the origin of parental care and the early evolution of male and female parental roles in birds. *The American Naturalist*, 143:39–58.

298 Cockburn, A. 2006. Prevalence of different modes of parental care in birds. *Proceedings of the Royal Society B: Biological Sciences*, 273:1375–1383.

299 Wade, M.J., and Shuster, S.M. 2002. The evolution of parental care in the context of sexual selection: a critical reassessment of parental investment theory. *The American Naturalist*, 160:285–292.

300 Varricchio, D.J., Moore, J.R., Erickson, G.M., Norell, M.A., Jackson, F.D., and Borkowski, J.J. 2008. Avian paternal care had dinosaur origin. *Science*, 322:1826–1828.

301 Liker, A., Freckleton, R.P., Remeš, V., and Székely, T. 2015. Sex differences in parental care: gametic investment, sexual selection, and social environment. *Evolution*, 69:2862–2875.

302 Prondvai, E. 2017. Medullary bone in fossils: function, evolution and significance in growth curve reconstructions of extinct vertebrates. *Journal of Evolutionary Biology*, 30:440–460.

303 Uomini, N., Fairlie, J., Gray, R. D., and Griesser, M. 2020. Extended parenting and the evolution of cognition. *Philosophical Transactions of the Royal Society B: Biological Sciences*, 375:20190495.

304 Meng, Q., Liu, J., Varricchio, D.J., Huang, T., and Gao, C. 2004. Palaeontology: Parental care in an ornithischian dinosaur. *Nature*, 431:145.

305 Varricchio, D.J., Martin, A.J. and Katsura, Y. 2007. First trace and body fossil evidence of a burrowing, denning dinosaur. *Proceedings of the Royal Society B: Biological Sciences*, 274:1361–1368.

306 Varricchio, D.J. 2010. A distinct dinosaur life history? *Historical Biology*, 23:91–107.

307 Lockley, M.G. 1990. Dinosaur ontogeny and population structure: interpretations and speculations based on fossil footprints. In: *Dinosaur Systematics: Approaches and Perspectives*, 347–365. Cambridge University Press.

308 Horner, J.R. 1984. The nesting behavior of dinosaurs. *Scientific American*, 250:130–137.

309 Horner, J.R. 1982. Evidence of colonial nesting and 'site fidelity' among ornithischian dinosaurs. *Nature*, 297:675–676.

310 Varricchio, D. J., and Horner, J.R. 1993. Hadrosaurid and lambeosaurid bone beds from the Upper Cretaceous Two Medicine Formation of Montana: taphonomic and biologic implications. *Canadian Journal of Earth Sciences*, 30:997–1006.

311 Hardy, I.C., and Briffa, M., eds. 2013. *Animal Contests*. Cambridge University Press.

312 Emlen, D.J. 2008. The evolution of animal weapons. *Annual Review of Ecology, Evolution, and Systematics*, 39:387–413.

313 Eibl-Eibesfeldt, I. 1961. The fighting behavior of animals. *Scientific American*, 205:112–123.

314 McCullough, E.L., Tobalske, B.W., and Emlen, D.J. 2014. Structural adaptations to diverse fighting styles in sexually selected weapons. *Proceedings of the National Academy of Sciences*, 111:14484–14488.

315 Gerald, M.S. 2001. Primate colour predicts social status and aggressive outcome. *Animal Behaviour*, 61:559–566.

316 Wood, K.A., Ponting, J., D'Costa, N., Newth, J.L., Rose, P.E., Glazov, P., and Rees, E.C. 2017. Understanding intrinsic and extrinsic drivers of aggressive behaviour in waterbird assemblages: a meta-analysis. *Animal Behaviour*, 126:209–216.

317 Matsumoto, K., and Knell, R.J. 2017. Diverse and complex male polymorphisms in *Odontolabis* stag beetles (Coleoptera: Lucanidae). *Scientific Reports*, 7:1–11.

318 Stankowich, T., and Caro, T. 2009. Evolution of weaponry in female bovids. *Proceedings of the Royal Society B: Biological Sciences*, 276:4329–4334.

319 Caro, T.M., Graham, C.M., Stoner, C.J., and Flores, M.M. 2003. Correlates of horn and antler shape in bovids and cervids. *Behavioral Ecology and Sociobiology*, 55:32–41.

320 Bro-Jørgensen, J. 2007. The intensity of sexual selection predicts weapon size in male bovids. *Evolution: International Journal of Organic Evolution*, 61:1316–1326.

321 Farke, A.A. 2014. Evaluating combat in ornithischian dinosaurs. *Journal of Zoology*, 292:242–249.

322 Remes, K., Ortega, F., Fierro, I., Joger, U., Kosma, R., Marín Ferrer, J.M., Ide, O.A., and Maga, A. 2009. A new basal sauropod dinosaur from the Middle Jurassic of Niger and the early evolution of Sauropoda. *PLoS One*, 4:e6924.

323 Soto-Acuña, S., Vargas, A.O., Kaluza, J., Leppe, M.A., Botelho, J.F., Palma-Liberona, J., Simon-Gutstein, C., Fernández, R.A., Ortiz, H., Milla, V., and Aravena, B. 2021. Bizarre tail weaponry in a transitional ankylosaur from subantarctic Chile. *Nature*, 600:259–263.

324 Farlow, J.O., and Dodson, P. 1975. The behavioral significance of frill and horn morphology in ceratopsian dinosaurs. *Evolution*, 29:353–361.

325 Farke, A.A. 2004. Horn use in *Triceratops* (Dinosauria: Ceratopsidae): testing behavioral hypotheses using scale models. *Palaeontologica Electronica*, 7:10.

326 Lautenschlager, S. 2014. Morphological and functional diversity in therizinosaur claws and the implications for theropod claw evolution. *Proceedings of the Royal Society B: Biological Sciences*, 281:20140497.

327 Hone, D.W.E., and Tanke, D.H. 2015. Pre- and postmortem tyrannosaurid bite marks on the remains of *Daspletosaurus* (Tyrannosaurinae: Theropoda) from Dinosaur Provincial Park, Alberta, Canada. *PeerJ*, 3:e885.

328 Galton, P.M. 1971. A primitive dome-headed dinosaur (Ornithischia: Pachy-cephalosauridae) from the Lower Cretaceous of England and the function of the dome of pachycephalosaurids. *Journal of Paleontology*, 45:40–47.

329 Snively, E., and Cox, A. 2008. Structural mechanics of pachycephalosaur crania permitted head-butting behavior. *Palaeontologia Electronica*, 11:3A.

330 Snively, E., and Theodor, J.M. 2011. Common functional correlates of head-strike behavior in the pachycephalosaur *Stegoceras validum* (Ornithischia, Dinosauria) and combative artiodactyls. *PLoS One*, 6:e21422.

331 Peterson, J.E., and C.P. Vittore. 2012. Cranial pathologies in a specimen of *Pachy-cephalosaurus*. *PloS One*, 7:e36227.

332 Ackermans, N.L., Varghese, M., Williams, T.M., Grimaldi, N., Selmanovic, E., Alipour, A., Balchandani, P., Reidenberg, J.S., and Hof, P.R. 2022. Evidence of traumatic brain injury in headbutting bovids. *Acta Neuropathologica*, 144:5–26.

333 Fastovsky, D.E., and Weishampel, D.B., 2013. *Dinosaurs: A Concise Natural History*, 2nd ed. Cambridge University Press.

334 Sues, H. 1978. Functional morphology of the dome in pachycephalosaurid dino-saurs. *Neues Jahrbuch für Geologie und Paläontologie*, Monatsheft. 1978:459–472.

335 Arbour, V.M., and Zanno, L.E. 2020. Tail weaponry in ankylosaurs and glypt-odonts: an example of a rare but strongly convergent phenotype. *The Anatomical Record*, 303:988–998.

336 Arbour, V.M. 2009. Estimating Impact Forces of Tail Club Strikes by Ankylo-saurid Dinosaurs. *PLoS One*, 4:e6738.

337 Carpenter, K., Sanders, F., McWhinney, L.A., and Wood, L. 2005. Evidence for predator-prey relationships: examples for *Allosaurus* and *Stegosaurus*. In: *The Carnivorous Dinosaurs*, 325–350. Indiana University Press, Bloomington.

338 Arbour, V.M. and Currie, P.J. 2011. Tail and pelvis pathologies of ankylosaurian dinosaurs. *Historical Biology*, 23:375–390.

339 Blows, W.T. 2001. Dermal armor of the polacanthine dinosaurs. In: *The Armored Dinosaurs*, 363–385. Indiana University Press, Bloomington.

340 Cau, A., Dalla Vecchia, F.M., and Fabbri, M. 2013. A thick-skulled theropod (Di-nosauria, Saurischia) from the Upper Cretaceous of Morocco with implications for carcharodontosaurid cranial evolution. *Cretaceous Research*, 40:251–260.

341 Mazzetta, G.V., Cisilino, A.P., Blanco, R.E., and Calvo, N. 2009. Cranial mechan-ics and functional interpretation of the horned carnivorous dinosaur *Carnotaurus sastrei*. *Journal of Vertebrate Paleontology*, 29:822–830.

342 Taylor, M.P., Wedel, M.J., Naish, D., and Engh, B. 2015. Were the necks of *Apatosaurus* and *Brontosaurus* adapted for combat? *PeerJ PrePrints*.

343 Lida, X., Yong, Y.E., Chunkang, S.H.U., Guangzhao, P., and Hailu, Y. 2009. Structure, orientation and finite element analysis of the tail club of *Mamenchisaurus hochuanensis*. *Acta Geologica Sinica* (English ed.), 83:1031–1040.

344 Arbour, V.M., Zanno, L.E., and Evans, D.C. 2022. Palaeopathological evidence for intraspecific combat in ankylosaurid dinosaurs. *Biology Letters*, 18:20220404.

345 Holtz, T.R. 2003. Dinosaur predation: evidence and ecomorphology. In: *Predator–Prey Interactions in the Fossil Record*, 325–340. Springer, New York.

346 McArthur, C., Banks, P.B., Boonstra, R., and Forbey, J.S. 2014. The dilemma of foraging herbivores: dealing with food and fear. *Oecologia*, 176:677–689.

347 Schmidt-Nielsen, K. 1984. *Scaling: Why Is Animal Size So Important?*. Cambridge University Press.

348 Janis, C.M., and Ehrhardt, D. 1988. Correlation of relative muzzle width and relative incisor width with dietary preference in ungulates. *Zoological Journal of the Linnean Society*, 92:267–284.

349 Krause, J., and Godin, J.G.J. 1996. Influence of prey foraging posture on flight behavior and predation risk: predators take advantage of unwary prey. *Behavioral Ecology*, 7:264–271.

350 Costa, G.C. 2009. Predator size, prey size, and dietary niche breadth relationships in marine predators. *Ecology*, 90:2014–2019.

351 Hirt, M.R., Tucker, M., Müller, T., Rosenbaum, B., and Brose, U. 2020. Rethinking trophic niches: speed and body mass colimit prey space of mammalian predators. *Ecology and Evolution*, 10:7094–7105.

352 Holekamp, K.E., Smale, L., Berg, R., and Cooper, S.M. 1997. Hunting rates and hunting success in the spotted hyena (*Crocuta crocuta*). *Journal of Zoology*, 242:1–15.

353 Schnitzler, A. and Hermann, L. 2019. Chronological distribution of the tiger *Panthera tigris* and the Asiatic lion *Panthera leo persica* in their common range in Asia. *Mammal Review*, 49:340–353.

354 Vandermeer, J.H. 1972. Niche theory. *Annual Review of Ecology and Systematics*, 3:107–132.

355 Kruuk, H., and Turner, M. 1967. Comparative notes on predation by lion, leopard, cheetah and wild dog in the Serengeti area, East Africa. *Mammalia*, 31:1–27

356 Mallon, J.C., Evans, D.C., Ryan, M.J., and Anderson, J.S. 2013. Feeding height stratification among the herbivorous dinosaurs from the Dinosaur Park Formation (upper Campanian) of Alberta, Canada. *BMC Ecology*, 13:1–15.

357 Ősi, A., Prondvai, E., Mallon, J., and Bodor, E.R. 2017. Diversity and convergences in the evolution of feeding adaptations in ankylosaurs (Dinosauria: Ornithischia). *Historical Biology*, 29:539–570.

358 Takasaki, R., and Kobayashi, Y. 2022. Beak morphology and limb proportions as adaptations of hadrosaurid foraging ecology. *Cretaceous Research*, 141:105361.

359 Mallon, J.C., and Anderson, J.S. 2014. The functional and palaeoecological implications of tooth morphology and wear for the megaherbivorous dinosaurs from the Dinosaur Park Formation (Upper Campanian) of Alberta, Canada. *PloS One*, 9:e98605.

360 Brown, C.M., Greenwood, D.R., Kalyniuk, J.E., Braman, D.R., Henderson, D.M., Greenwood, C.L., and Basinger, J.F. 2020. Dietary palaeoecology of an Early Cretaceous armoured dinosaur (Ornithischia; Nodosauridae) based on floral analysis of stomach contents. *Royal Society Open Science*, 7:200305.

361 Chin, K., Feldmann, R.M., and Tashman, J.N. 2017. Consumption of crustaceans by megaherbivorous dinosaurs: dietary flexibility and dinosaur life history strategies. *Scientific Reports*, 7:1–11.

362 Hummel, J., Gee, C.T., Südekum, K.H., Sander, P.M., Nogge, G., and Clauss, M. 2008. In vitro digestibility of fern and gymnosperm foliage: implications for sauropod feeding ecology and diet selection. *Proceedings of the Royal Society B: Biological Sciences*, 275:1015–1021.

363 Button, D.J., Rayfield, E.J., and Barrett, P.M. 2014. Cranial biomechanics underpins high sauropod diversity in resource-poor environments. *Proceedings of the Royal Society B: Biological Sciences*, 281:20142114.

364 Woodruff, D.C., Carr, T.D., Storrs, G.W., Waskow, K., Scannella, J.B., Nordén, K.K., and Wilson, J.P. 2018. The smallest diplodocid skull reveals cranial ontogeny and growth-related dietary changes in the largest dinosaurs. *Scientific Reports*, 8:1–12.

365 Lautenschlager, S., Brassey, C.A., Button, D.J., and Barrett, P.M. 2016. Decoupled form and function in disparate herbivorous dinosaur clades. *Scientific Reports*, 6:1–10.

366 Ma, W.S., Pittman, M., Butler, R. and Lautenschlager, S. 2021. Macroevolutionary trends in theropod dinosaur feeding mechanics. *Current Biology*, 32:677–686.

367 Meade, L.E., and Ma, W. 2022. Cranial muscle reconstructions quantify adaptation for high bite forces in Oviraptorosauria. *Scientific Reports*, 12:1–15.

368 Barrett, P.M. 2005. The diet of ostrich dinosaurs (Theropoda: Ornithomimosauria). *Palaeontology*, 48:347–358.

369 Ji, Q., Currie, P.J., Norell, M.A., and Shu-An, J. 1998. Two feathered dinosaurs from northeastern China. *Nature*, 393:753–761.

370 Kobayashi, Y., Lu, J.C., Dong, Z.M., Barsbold, R., Azuma, Y., and Tomida, Y. 1999. Herbivorous diet in an ornithomimid dinosaur. *Nature*, 402:480–481.

371 Lee, Y.N., Barsbold, R., Currie, P.J., Kobayashi, Y., Lee, H.J., Godefroit, P., Escuillié, F., and Chinzorig, T. 2014. Resolving the long-standing enigmas of a giant ornithomimosaur *Deinocheirus mirificus*. *Nature*, 515:257–260.

372 Hone, D.W.E., Dececchi, T.A., Sullivan, C., Xu, X., and Larsson, H.C.E. 2023. Generalist diet of *Microraptor zhaoianus* included mammals. *Journal of Vertebrate Paleontology*, 42:e2144337.

373 Holtz, T.R. 2008. A critical reappraisal of the obligate scavenging hypothesis for *Tyrannosaurus rex* and other tyrant dinosaurs. In: Tyrannosaurus rex, *the Tyrant King*, 371–396. Indiana University Press, Bloomington.

374 Dal Sasso, C., and Maganuco, S. 2011. *Scipionyx samniticus* (Theropoda: Compsognathidae) from the Lower Cretaceous of Italy. Osteology, ontogenetic assessment, phylogeny, soft tissue anatomy, taphonomy and paleobiology. Mono-

graph. *Società Italiana di Scienze Naturali e Museo di Storia Naturale di Milano*, 37:1–282.

375 Jacobsen, A.R. 1998. Feeding behaviour of carnivorous dinosaurs as determined by tooth marks on dinosaur bones. *Historical Biology*, 13:17–26.

376 Sakamoto, M. 2022. Estimating bite force in extinct dinosaurs using phylogenetically predicted physiological cross-sectional areas of jaw adductor muscles. *PeerJ*, 10:e13731.

377 Erickson, G.M., Van Kirk, S.D., Su, J., Levenston, M.E., Caler, W.E., and Carter, D.R. 1996. Bite-force estimation for *Tyrannosaurus rex* from tooth-marked bones. *Nature*, 382: 706.

378 Powers, M.J., Sullivan, C., and Currie, P.J. 2020. Re-examining ratio based premaxillary and maxillary characters in Eudromaeosauria (Dinosauria: Theropoda): Divergent trends in snout morphology between Asian and North American taxa. *Palaeogeography, Palaeoclimatology, Palaeoecology*, 547:109704.

379 Blumenschine, R.J. 1987. Characteristics of an early hominid scavenging niche. *Current Anthropology*, 28:383–417.

380 Snively, E., and Russell, A.P. 2007. Functional variation of neck muscles and their relation to feeding style in Tyrannosauridae and other large theropod dinosaurs. *The Anatomical Record*, 290:934–957.

381 Fowler, D.W., Freedman, E.A., Scannella, J.B., and Kambic, R.E. 2011. The predatory ecology of *Deinonychus* and the origin of flapping in birds. *PLoS One*, 6:e28964.

382 Bishop, P.J. 2019. Testing the function of dromaeosaurid (Dinosauria, Theropoda) 'sickle claws' through musculoskeletal modelling and optimization. *PeerJ*, 7:e7577.

383 Pittman, M., Bell, P.R., Miller, C.V., Enriquez, N.J., Wang, X., Zheng, X., Tsang, L.R., Tse, Y.T., Landes, M., and Kaye, T.G. 2022. Exceptional preservation and foot structure reveal ecological transitions and lifestyles of early theropod flyers. *Nature Communications*, 13:7684.

384 Oswald, T., Curtice, B., Bolander, M., and Lopez, C. 2023. Observation of claw use and feeding behavior of the red-legged seriema and its implication for claw use in deinonychosaurs. *Journal of the Arizona-Nevada Academy of Science*, 50:17–21.

385 Persons IV, W.S., and Currie, P.J. 2011. Dinosaur speed demon: the caudal musculature of *Carnotaurus sastrei* and implications for the evolution of South American abelisaurids. *PloS One*, 6:e25763.

386 Persons, W.S., Currie, P.J., Eberth, D.A., and Evans, D.C. 2014. Duckbills on the run: the cursorial abilities of hadrosaurs and implications for tyrannosaur-avoidance strategies. In: *Hadrosaurs*, 449–458). Indiana University Press, Bloomington.

387 Qin, Z., Liao, C.C., Benton, M.J., and Rayfield, E.J. 2023. Functional space analyses reveal the function and evolution of the most bizarre theropod manual unguals. *Communications Biology*, 6:181.

388 Frederickson, J.A., Engel, M.H. and Cifell, R.L. 2020. Ontogenetic dietary shifts in *Deinonychus antirrhopus* (Theropoda; Dromaeosauridae): Insights into the ecology and social behavior of raptorial dinosaurs through stable isotope analysis. *Palaeogeography, Palaeoclimatology, Palaeoecology*, 552:109780.

389 Seymour, R.S. 2016. Cardiovascular physiology of dinosaurs. *Physiology*, 31:430–441.

390 Stevens, K.A. 2013. The articulation of sauropod necks: methodology and mythology. *PLoS One*, 8:e78572.

391 Stevens, K.A., and Parrish, J.M. 2005. Neck posture, dentition, and feeding strategies in Jurassic sauropod dinosaurs. In: *Thunder-Lizards: The Sauropodomorph Dinosaurs*, eds. V. Tidwell and K. Carpenter, 212–232. Indiana University Press, Bloomington.

392 Woodruff, D.C. 2017. Nuchal ligament reconstructions in diplodocid sauropods support horizontal neck feeding postures. *Historical Biology*, 29:308–319.

393 Prasad, V., Stromberg, C.A., Alimohammadian, H., and Sahni, A. 2005. Dinosaur coprolites and the early evolution of grasses and grazers. *Science*, 310(5751):1177–1180.

394 Barrett, P.M., and Upchurch, P. 1994. Feeding mechanisms of *Diplodocus*. *Gaia: Revista de Geociências, Museu Nacional de História Natural*, 10:195–204.

395 Young, M.T., Rayfield, E.J., Holliday, C.M., Witmer, L.M., Button, D.J., Upchurch, P., and Barrett, P.M. 2012. Cranial biomechanics of *Diplodocus* (Dinosauria, Sauropoda): testing hypotheses of feeding behaviour in an extinct megaherbivore. *Naturwissenschaften*, 99:637–643.

396 Whitlock, J.A. 2011. Inferences of diplodocoid (Sauropoda: Dinosauria) feeding behavior from snout shape and microwear analyses. *PLoS One*, 6:e18304.

397 Price, J.R., and Whitlock, J.A. 2022. Dental histology of *Diplodocus* (Sauropoda, Diplodocoidea). *Journal of Vertebrate Paleontology*, 42:e2099745.

398 Cameron, E.Z., and du Toit, J.T. 2007. Winning by a neck: tall giraffes avoid competing with shorter browsers. *The American Naturalist*, 169:130–135.

399 Vidal, D., Mocho, P., Aberasturi, A., Sanz, J.L., and Ortega, F. 2020. High browsing skeletal adaptations in *Spinophorosaurus* reveal an evolutionary innovation in sauropod dinosaurs. *Scientific Reports*, 10:1–10.

400 Falkingham, P.L., Bates, K.T., and Farlow, J.O. 2014. Historical photogrammetry: Bird's Paluxy River dinosaur chase sequence digitally reconstructed as it was prior to excavation 70 years ago. *PloS One*, 9:e93247.

INDEX

Page numbers in italics indicate figures and tables.